LONDON MATHEMATICAL SOCIETY LECTURE NOTE SERIES

Managing Editor: Professor I.M. James,
Mathematical Institute, 24-29 St Giles,Oxford

1. General cohomology theory and K-theory, P.HILTON
4. Algebraic topology, J.F.ADAMS
5. Commutative algebra, J.T.KNIGHT
8. Integration and harmonic analysis on compact groups, R.E.EDWARDS
9. Elliptic functions and elliptic curves, P.DU VAL
10. Numerical ranges II, F.F.BONSALL & J.DUNCAN
11. New developments in topology, G.SEGAL (ed.)
12. Symposium on complex analysis, Canterbury, 1973, J.CLUNIE & W.K.HAYMAN (eds.)
13. Combinatorics: Proceedings of the British Combinatorial Conference 1973, T.P.McDONOUGH & V.C.MAVRON (eds.)
15. An introduction to topological groups, P.J.HIGGINS
16. Topics in finite groups, T.M.GAGEN
17. Differential germs and catastrophes, Th.BROCKER & L.LANDER
18. A geometric approach to homology theory, S.BUONCRISTIANO, C.P. ROURKE & B.J.SANDERSON
20. Sheaf theory, B.R.TENNISON
21. Automatic continuity of linear operators, A.M.SINCLAIR
23. Parallelisms of complete designs, P.J.CAMERON
24. The topology of Stiefel manifolds, I.M.JAMES
25. Lie groups and compact groups, J.F.PRICE
26. Transformation groups: Proceedings of the conference in the University of Newcastle-upon-Tyne, August 1976, C.KOSNIOWSKI
27. Skew field constructions, P.M.COHN
28. Brownian motion, Hardy spaces and bounded mean oscillations, K.E.PETERSEN
29. Pontryagin duality and the structure of locally compact Abelian groups, S.A.MORRIS
30. Interaction models, N.L.BIGGS
31. Continuous crossed products and type III von Neumann algebras, A.VAN DAELE
32. Uniform algebras and Jensen measures, T.W.GAMELIN
33. Permutation groups and combinatorial structures, N.L.BIGGS & A.T.WHITE
34. Representation theory of Lie groups, M.F. ATIYAH et al.
35. Trace ideals and their applications, B.SIMON
36. Homological group theory, C.T.C.WALL (ed.)
37. Partially ordered rings and semi-algebraic geometry, G.W.BRUMFIEL
38. Surveys in combinatorics, B.BOLLOBAS (ed.)
39. Affine sets and affine groups, D.G.NORTHCOTT
40. Introduction to Hp spaces, P.J.KOOSIS
41. Theory and applications of Hopf bifurcation, B.D.HASSARD, N.D.KAZARINOFF & Y-H.WAN
42. Topics in the theory of group presentations, D.L.JOHNSON
43. Graphs, codes and designs, P.J.CAMERON & J.H.VAN LINT
44. Z/2-homotopy theory, M.C.CRABB
45. Recursion theory: its generalisations and applications, F.R.DRAKE & S.S.WAINER (eds.)
46. p-adic analysis: a short course on recent work, N.KOBLITZ
47. Coding the Universe, A.BELLER, R.JENSEN & P.WELCH
48. Low-dimensional topology, R.BROWN & T.L.THICKSTUN (eds.)

49. Finite geometries and designs, P.CAMERON, J.W.P.HIRSCHFELD & D.R.HUGHES (eds.)
50. Commutator calculus and groups of homotopy classes, H.J.BAUES
51. Synthetic differential geometry, A.KOCK
52. Combinatorics, H.N.V.TEMPERLEY (ed.)
53. Singularity theory, V.I.ARNOLD
54. Markov processes and related problems of analysis, E.B.DYNKIN
55. Ordered permutation groups, A.M.W.GLASS
56. Journées arithmétiques 1980, J.V.ARMITAGE (ed.)
57. Techniques of geometric topology, R.A.FENN
58. Singularities of smooth functions and maps, J.MARTINET
59. Applicable differential geometry, M.CRAMPIN & F.A.E.PIRANI
60. Integrable systems, S.P.NOVIKOV et al.
61. The core model, A.DODD
62. Economics for mathematicians, J.W.S.CASSELS
63. Continuous semigroups in Banach algebras, A.M.SINCLAIR
64. Basic concepts of enriched category theory, G.M.KELLY
65. Several complex variables and complex manifolds I, M.J.FIELD
66. Several complex variables and complex manifolds II, M.J.FIELD
67. Classification problems in ergodic theory, W.PARRY & S.TUNCEL
68. Complex algebraic surfaces, A.BEAUVILLE
69. Representation theory, I.M.GELFAND et al.
70. Stochastic differential equations on manifolds, K.D.ELWORTHY
71. Groups - St Andrews 1981, C.M.CAMPBELL & E.F.ROBERTSON (eds.)
72. Commutative algebra: Durham 1981, R.Y.SHARP (ed.)
73. Riemann surfaces: a view towards several complex variables, A.T.HUCKLEBERRY
74. Symmetric designs: an algebraic approach, E.S.LANDER
75. New geometric splittings of classical knots (algebraic knots), L.SIEBENMANN & F.BONAHON
76. Linear differential operators, H.O.CORDES
77. Isolated singular points on complete intersections, E.J.N.LOOIJENGA
78. A primer on Riemann surfaces, A.F.BEARDON
79. Probability, statistics and analysis, J.F.C.KINGMAN & G.E.H.REUTER (eds.)
80. Introduction to the representation theory of compact and locally compact groups, A.ROBERT
81. Skew fields, P.K.DRAXL
82. Surveys in combinatorics: Invited papers for the ninth British Combinatorial Conference 1983, E.K.LLOYD (ed.)
83. Homogeneous structures on Riemannian manifolds, F.TRICERRI & L.VANHECKE
84. Finite group algebras and their modules, P.LANDROCK
85. Solitons, P.G.DRAZIN
86. Topological topics, I.M.JAMES (ed.)
87. Surveys in set theory, A.R.D.MATHIAS (ed.)

London Mathematical Society Lecture Note Series: 87

Surveys in set theory

Edited by A.R.D. Mathias
Fellow of Peterhouse, Cambridge

CAMBRIDGE UNIVERSITY PRESS

Cambridge

London New York New Rochelle

Melbourne Sydney

CAMBRIDGE UNIVERSITY PRESS
Cambridge, New York, Melbourne, Madrid, Cape Town, Singapore, São Paulo

Cambridge University Press
The Edinburgh Building, Cambridge CB2 8RU, UK

Published in the United States of America by Cambridge University Press, New York

www.cambridge.org
Information on this title: www.cambridge.org/9780521277334

First published 1983
Re-issued in this digitally printed version 2008

A catalogue record for this publication is available from the British Library

Library of Congress Catalogue Card Number: 83–10106

ISBN 978-0-521-27733-4 paperback

CONTENTS

PREFACE

In 1978 a Summer School in set theory of three weeks' duration was held in Cambridge, England, and was attended by about one hundred people. It was a most stimulating meeting, and the Organising Committee, Professor L. Harrington, Professor M. Magidor and the undersigned, here record their gratitude to the Scientific Affairs Division of NATO, who funded the meeting as an Advanced Study Institute; the Master and Fellows of Peterhouse; the Master and Fellows of Pembroke College; the University of Cambridge; and in particular to Mr and Mrs J. F. Jenkins who ensured the smooth running of the whole. The names of participants with their addresses in 1978 are recorded at the end of the volume.

In some cases lecturers considered that adequate accounts of their topics were already in print or in press. This volume comprises five expository papers: those by Baumgartner, Guaspari and Stanley are developments of lectures presented at the meeting, those by Devlin and Kanamori, which expound topics complementing those of Baumgartner and Stanley respectively, were written independently of the meeting; and two research papers: those by Prikry and Shelah.

A.R.D. Mathias

Iterated forcing

James E. Baumgartner[1]

0. Introduction and terminology.

In view of the great number of independence problems which have been solved by Cohen's method of forcing, it is not surprising that the ones remaining seem to require more and more sophisticated techniques for their solution. One such technique receiving considerable attention recently is the technique of *iterated forcing*, in which the set of forcing conditions is built up inductively and the ultimate generic extension is regarded as having been obtained in a series of stages, each one generic over the ones preceding. This technique has its roots in the Solovay-Tennenbaum consistency proof [21] of Martin's Axiom, but recently there has been a series of striking applications of the method and its variations such as Silver's use of reverse Easton forcing, Laver's proof [11] of the consistency of the Borel Conjecture, and Shelah's consistency proof [16] of the nonexistence of P-point ultrafilters.

This paper is an introduction to iterated forcing for the reader already familiar with the rudiments of forcing. A more complete discussion of prerequisites will be found below.

The paper is organized as follows. Section 1 contains the general definition of an α-stage iteration, together with some

---

[1] Research for this paper was partially supported by National Science Foundation grants MCS76-08231A01 and MCS79-03376.

elementary observations about iterated forcing. In Section 2 the basic facts about chain conditions and closure conditions in iterations are proved, and these results are immediately applied in Sections 3 and 4 to obtain the consistency of Martin's Axiom and of a generalized form of Martin's Axiom.

With the experience acquired in Sections 3 and 4, we pause in Section 5 to look at intermediate stages in iterated extensions. Section 6 gives an example of the method of reverse Easton forcing, used by Silver in unpublished work (see [13]) to solve the continuum problem for an extensive class of large cardinals. The result presented is a weak version of one of Silver's results, namely if κ is κ^{++}-supercompact and $2^\kappa = \kappa^+$ then it is possible to find a generic extension in which κ is still measurable and $2^\kappa = \kappa^{++}$.

In Section 7 a property of partial orderings, called Axiom A, is described which applies both to countable chain condition orderings and countably closed orderings, as well as to other well-known forcing methods for adding real numbers. It is shown that iterated Axiom A forcing does not collapse ω_1, and that under certain circumstances the iteration has the $\aleph_2$-chain condition. These results are applied in Section 8 to give a relatively straightforward consistency proof for Martin's Axiom together with the assertion that every tree of height ω_1 and cardinality $\aleph_1$ has at most $\aleph_1$ uncountable branches.

Finally, in Section 9 we use the results of Section 7 to give an alternate proof for the consistency of the Borel Conjecture using iterated Mathias forcing. This approach was suggested by a remark of Laver.

The author wishes to express his thanks to Richard Laver and Saharon Shelah for valuable conversations and correspondence. The approach to iterated forcing taken here was strongly influenced by Laver's paper [11].

Thanks are also due to Stephen Simpson and Alan Taylor for having read and commented upon preliminary versions of this paper.

Our forcing terminology is as follows. If $(P, \leq)$ is a partial ordering and $p,q \in P$ then the relation $p \leq q$ is read as p <u>extends</u> q, or p is <u>stronger than</u> q. (Some authors write this as $p \geq q$.) We say p and q are <u>compatible</u> if $(\exists r \in P)\ r \leq p,q$; otherwise p and q are <u>incompatible</u>. A set $D \subseteq P$ is <u>dense in</u> P iff $(\forall p \in P)(\exists q \in D)\ q \leq p$. We assume that every partial ordering has a maximum element, which will be denoted by 1 (thus $p \leq 1$ for all $p \in P$).

We frequently use preorderings (i.e., sets bearing a reflexive, transitive relation) as sets of forcing conditions. When this is done, we always identify p and q if it is true that $p \leq q$ and $q \leq p$. We are actually passing to the partial ordering of equivalence classes, but it is more convenient to avoid the awkward equivalence class notation.

Forcing is always considered to be taking place over V, the universe of all sets, or some transitive class model of ZFC. The symbol $\Vdash_P$ denotes (weak) forcing with respect to P. Since in iterated forcing subscripts on partial orderings frequently occur, we will use $\Vdash_\alpha$ as an abbreviation for $\Vdash_{P_\alpha}$. If ϕ is a sentence of the language of forcing, then $\Vdash_P \phi$ means $(\forall p \in P)\ p \Vdash_P \phi$ (or $1 \Vdash_P \phi$).

A set $G \subseteq P$ is P-generic over a class V' iff

(a) $(\forall p,q \in G)(\exists r \in G)\ r \leq p,q$

(b) $(\forall p \in G)(\forall q \in P)$ if $p \leq q$ then $q \in G$

(c) if $D \in V'$ and D is dense in P then $G \cap D \neq 0$.

If G is P-generic over V', then V'[G] denotes the generic extension.

The reader is presumed to be familiar with one of the modern developments of forcing for arbitrary partial orderings (such as [9] or [18] but not [3]). In particular it will be assumed that the reader knows the fundamental theorems about forcing (such as that $p \Vdash_P \neg\phi$ iff $(\forall q \leq p)$ it is not the case that $q \Vdash_P \phi$) as well as the effects of chain conditions and closure conditions on forcing arguments.

An antichain in a partial ordering P is a pairwise incompatible set. If κ is a cardinal, then P has the κ-chain condition iff for any antichain $A \subseteq P$, $|A| < \kappa$. P is κ-closed iff for any $\lambda < \kappa$ and any sequence $\langle p_\alpha : \alpha < \lambda\rangle$ such that $\alpha < \beta$ implies $p_\beta \leq p_\alpha$, there is $p \in P$ such that $p \leq p_\alpha$ for all $\alpha < \lambda$. The expressions "countable chain condition" and "countably closed" are generally used instead of "$\aleph_1$-chain condition" and "$\aleph_1$-closed". We say P is κ-directed closed iff for any set $X \subseteq P$, if X is directed (i.e., if $(\forall p,q \in X)(\exists r \in X)\ r \leq p,q$) and $|X| < \kappa$ then there exists $p \in P$ such that $(\forall q \in X)\ p \leq q$. Note that if P is κ-directed closed then P is κ-closed, and if $\kappa = \aleph_1$ then the converse is true as well. It is well-known that if P has the κ-chain condition then forcing P preserves all cardinals $\geq \kappa$, while if P is κ-closed then forcing with P preserves all cardinals $\leq \kappa$.

A family of sets F is a Δ-system iff there is a set Δ, called the kernel of the Δ-system, such that for all $A,B \in F$, if $A \neq B$ then $A \cap B = \Delta$. The κ-chain condition is often verified using Δ-systems.

If $x \in V$ then we will use x as a term of the language of forcing over V to denote x. Some autgors use $\check{x}$ to denote x; our choice is simply a matter of typographical convenience. Other terms, i.e., terms not necessarily denoting specific elements of

V, will be denoted by letters with a dot over the top. Thus, for example, it is perfectly possible to have p $\Vdash_P$ $\dot{a} \in V$ without having p $\Vdash_P$ $\dot{a} = a$ for any specific $a \in V$. Of course it is always true that if p $\Vdash_P$ $\dot{a} \in V$, then $(\forall q \leq p)(\exists r \leq q)(\exists a \in V)$ r $\Vdash_P$ $\dot{a} = a$. The term $\dot{1}$ will always be used to denote the maximum element of a partial ordering.

We shall assume that the class of terms of the language of forcing with P over V is <u>full</u>, meaning that if p $\Vdash_P$ $\exists x\phi(x)$, then there is a term $\dot{x}$ such that p $\Vdash_P$ $\phi(\dot{x})$. This assumption is crucial for many arguments, and is often used withour comment. It requires the axiom of choice in V, which is assumed throughout the paper.

Sentences of the language of forcing are generally expressed in mathematical English. If the sentence is more than a few words in length, then it is usually set off from the rest of the text by quotation marks, thus p $\Vdash_P$ "$\dot{f}$ is a function mapping $\dot{x}$ into $\dot{y}$".

Since there are several different definitions of the class of terms of the language of forcing, we assume no particular definition here. Rather we make use of the fact that every element of V[G] is definable in V[G] from G together with (finitely many) elements of V. Every definition corresponds to a term, and every term corresponds to a definition. Accordingly, when we wish to specify a term $\dot{x}$ of the language of forcing, we simply specify how the object to be denoted by $\dot{x}$ is to be defined from G and certain elements of V.

If $\dot{x}$ is a term, then the object denoted or represented by $\dot{x}$ in the generic extension V[G] is written $\dot{x}^{V[G]}$.

1. Iterated forcing.

It is well-known that a generic extension of a generic extension is expressible as a single generic extension. Suppose P is a partial ordering and $\Vdash_P \dot{Q}$ is a partial ordering. Let $P \otimes \dot{Q} = \{(p,\dot{q}): p \in P \text{ and } \Vdash_P \dot{q} \in \dot{Q}\}$. Let $(p_1,\dot{q}_1) \leq (p_2,\dot{q}_2)$ iff $p_1 \leq p_2$ and $p_1 \Vdash \dot{q}_1 \leq \dot{q}_2$. We identify elements $(p_1,\dot{q}_1)$ and $(p_2,\dot{q}_2)$ if $(p_1,\dot{q}_1) \leq (p_2,\dot{q}_2) \leq (p_1,\dot{q}_1)$. Then forcing with $P \otimes \dot{Q}$ accomplishes exactly the same thing as forcing first with P and then with $\dot{Q}$. More precisely:

Theorem 1.1. (a) Suppose G is P-generic over V and H is $\dot{Q}^{V[G]}$-generic over V[G]. Let $J = \{(p,\dot{q}) \in P \otimes \dot{Q}: p \in G \text{ and } \dot{q}^{V[G]} \in H\}$. Then J is $P \otimes Q$-generic over V. (J is sometimes denoted by $G \otimes H$.)

(b) Suppose J is $P \otimes \dot{Q}$-generic over V. Then $G = \{p \in P: \exists \dot{q}\ (p,\dot{q}) \in J\}$ is P-generic over V and $H = \{\dot{q}^{V[G]}: \exists p\ (p,\dot{q}) \in J\}$ is $\dot{Q}^{V[G]}$-generic over V[G].

Proof. In each case we check only the denseness condition.

(a) Suppose D is dense in $P \otimes \dot{Q}$. Define E in V[G] by letting $\dot{q}^{V[G]} \in E$ iff $\exists p \in G\ (p,\dot{q}) \in D$. Let $\dot{E}$ be the canonical name for E. We claim $\Vdash_P \dot{E}$ is dense in $\dot{Q}$. Suppose $p \in P$ and $\Vdash_P \dot{q} \in \dot{Q}$. Since D is dense there is $(p_1,\dot{q}_1) \in D$ such that $(p_1,\dot{q}_1) \leq (p,\dot{q})$. But clearly $p_1 \Vdash$ "$q_1 \in \dot{E}$ and $\dot{q}_1 \leq \dot{q}$". Thus $\{p \in P: p \Vdash \exists q \leq \dot{q}\ q \in \dot{E}\}$ is dense in P and we're done. Now choose $\dot{q}^{V[G]} \in E \cap H$ and let $p \in G$ be such that $p \Vdash \dot{q} \in \dot{E}$. Then $(p,\dot{q}) \in J \cap D$.

(b) If D is dense in P then $\{(p,\dot{q}) \in P \otimes \dot{Q}: p \in D\}$ is dense in $P \otimes \dot{Q}$. Thus $G \cap D \neq 0$ and G is P-generic over V. If E is dense in $\dot{Q}^{V[G]}$ then there is a name $\dot{E}$ so that $E = \dot{E}^{V[G]}$ and $\Vdash_P \dot{E}$ is dense in $\dot{Q}$. But then $\{(p,\dot{q}) \in P \otimes \dot{Q}: p \Vdash \dot{q} \in \dot{E}\}$ is dense in $P \otimes \dot{Q}$ and the proof is complete. □

Remarks. 1. We will occasionally declare that $(p,\dot{q}) \in P \otimes \dot{Q}$ when we have only checked that $p \Vdash \dot{q} \in \dot{Q}$. This is permissible since there is always a term $\dot{r}$ such that $\Vdash \dot{r} \in \dot{Q}$ and $p \Vdash \dot{q} = \dot{r}$.

2. If $Q \in V$ then $P \times Q$ (with the coordinatewise ordering) is dense in $P \otimes Q$, provided each $q \in Q$ is identified with its canonical name. It is easy to see that in Theorem 1.1(a) and (b), $J \cap (P \times Q) = G \times H$. Thus if G is P-generic over V and H is Q-generic over V[G] then by symmetry we obtain the well-known result that G is also P-generic over V[H].

Now that we have seen how to handle a two-stage iteration, a three-stage iteration presents no problem. We simply force with $(P \otimes \dot{Q}) \otimes \dot{R}$. In order to treat iterations uniformly, it is natural to consider elements of $(P \otimes \dot{Q}) \otimes \dot{R}$ not to be pairs $((p,\dot{q}),\dot{r})$, but sequences $(p,\dot{q},\dot{r})$ of length three. This leads ultimately to the following inductive definition, which works even for iteration into the transfinite.

Definition. Let $\alpha \geq 1$. A partial ordering P_α is an α-stage iteration iff P_α is a set of α-sequences satisfying the following conditions:

(a) If $\alpha = 1$ then there is a partial ordering Q_0 so that $p \in P$ iff $p(0) \in Q_0$ and $p \leq q$ iff $p(0) \leq q(0)$. So $P_1 \cong Q_0$.

(b) If $\alpha = \beta + 1$, $\beta \geq 1$, then $P_\beta = \{p|\beta : p \in P_\alpha\}$ is a β-stage iteration and there is $\dot{Q}_\beta$ such that $\Vdash_\beta \dot{Q}_\beta$ is a partial ordering, and $p \in P_\alpha$ iff $p|\beta \in P_\beta$ and $\Vdash_\beta p(\beta) \in \dot{Q}_\beta$. Moreover, $p \leq q$ iff $p|\beta \leq q|\beta$ and $p|\beta \Vdash_\beta p(\beta) \leq q(\beta)$. (Note by induction that the ordering on P_β is uniquely determined). Thus $P_\alpha \cong P_\beta \otimes \dot{Q}_\beta$.

(c) If α is a limit ordinal then $\forall\beta < \alpha$ $P_\beta = \{p|\beta : p \in P_\alpha\}$ is a β-stage iteration, and

(i) $\bar{1} \in P_\alpha$, where $\bar{1}(\gamma) = \dot{1}$ for all $\gamma < \alpha$ (recall that $\dot{1}$ is the maximal element of $\dot{Q}_\gamma$).

(ii) if $\beta < \alpha$, $p \in P_\alpha$, $q \in P_\beta$ and $q \le p|\beta$, then $r \in P_\alpha$, where $r|\beta = q$ and $r(\gamma) = p(\gamma)$ for $\beta \le \gamma < \alpha$.

(iii) for all $p,q \in P_\alpha$, $p \le q$ iff $\forall\beta < \alpha\ p|\beta \le q|\beta$.

It is easy to see by induction that if $p \in P_\alpha$ then $p(0) \in Q_0$ and $\forall\beta < \alpha\ \beta \ge 1 \Rightarrow \Vdash_\beta p(\beta) \in \dot{Q}_\beta$. If $p,q \in P_\alpha$ then $p \le q$ iff $p(0) \le q(0)$ and $\forall\beta < \alpha\ \beta \ge 1 \Rightarrow p|\beta \Vdash p(\beta) \le q(\beta)$. These observations will be used constantly.

Note that P_1 is completely determined by Q_0, and $P_{\beta+1}$ is completely determined by P_β and $\dot{Q}_\beta$. If α is a limit ordinal, however, then except in certain trivial cases P_α is not determined by $\langle P_\beta: \beta < \alpha\rangle$. There are several possibilities:

We say that P_α is the <u>direct limit</u> of $\langle P_\beta: \beta < \alpha\rangle$ if $p \in P_\alpha$ iff $\exists\beta < \alpha\ p|\beta \in P_\beta$ and $\forall\gamma\ \beta \le \gamma < \alpha \Rightarrow p(\gamma) = \dot{1}$. We say that P_α is the <u>inverse limit</u> of $\langle P_\beta: \beta < \alpha\rangle$ if $p \in P_\alpha$ iff $\forall\beta < \alpha\ p|\beta \in P_\beta$. Other kinds of limits are possible but less common. Perhaps the best known intermediate limit is Jensen's construction [5] for obtaining the consistency of Souslin's Hypothesis with the Continuum Hypothesis.

Thus an α-stage iteration P_α will be completely determined if we specify the initial ordering Q_0, the orderings $\dot{Q}_\beta$ to be used at stages $\beta + 1 \le \alpha$, and the type of limit to be taken at each limit ordinal $\delta \le \alpha$.

If $p \in P_\alpha$ then the <u>support</u> of p is defined by support(p) = $\{\beta < \alpha: p(\beta) \ne \dot{1}\}$. Note, for example, that if direct limits are taken at every $\beta \le \alpha$ such that $\mathrm{cf}\beta > \omega$ (and any kind of limit elsewhere) then the support of every $p \in P_\alpha$ is countable or finite.

Some authors replace our P_α by $\{p|\mathrm{support}(p): p \in P_\alpha\}$. This is obviously an equivalent approach.

Henceforth, when the notation P_α is used the tacit assumption is made that P_α is an α-stage iteration. When the notation

$\dot{Q}_\beta$ is employed in connection with P_α, it has the meaning in the definition above. Also, if the notation G_α is used, then it will be assumed that G_α is P_α-generic over V. If G_β and G_α are used together, and $\beta < \alpha$, then it is assumed in addition that $G_\beta = \{p|\beta: p \in G_\alpha\}$. This assumption is justified by the following theorem.

Theorem 1.2. If $\beta < \alpha$, G_α is P_α-generic over V and $G_\beta = \{p|\beta: p \in G_\alpha\}$, then G_β is P_β-generic over V.

Proof. Again we check only the denseness condition. Suppose $D \subseteq P_\beta$ is dense. For $p \in P_\beta$ let $\bar{p} \in P_\alpha$ be such that $\bar{p}|\beta = p$ and $\bar{p}(\gamma) = \dot{1}$ if $\beta \le \gamma < \alpha$. Let $\bar{D} = \{q \in P_\alpha: (\exists p \in D)\ q \le \bar{p}\}$. We claim $\bar{D}$ is dense in P_α. If $r \in P_\alpha$ then $r|\beta \in P_\beta$ so there is $p \in D$ such that $p \le r|\beta$. If $q \in P_\alpha$ is defined so that $q|\beta = p$ and $q(\gamma) = r(\gamma)$ for $\beta \le \gamma < \alpha$, then $q \le \bar{p}$ and $q \in \bar{D}$. Hence $\bar{D}$ is dense. Then $G_\alpha \cap \bar{D} \neq 0$ so there is $p \in D$ such that $\bar{p} \in G_\alpha$. But then $p \in G_\beta \cap D$. □

One final comment will save considerable time in the proofs to come later. Part (a) of the definition of α-stage iteration may be regarded as a special case of part (b) for $\beta = 0$. If we define $P_0 = \{0\}$ then Q_0 is a partial ordering in the (trivial) generic extension of the universe via forcing with P_0. This allows us to omit a step in several proofs.

2. Chain conditions and closure.

Here we investigate the preservation of chain conditions and closure conditions by iterations.

Throughout this section, κ denotes a regular uncountable cardinal.

Theorem 2.1. If P has the κ-chain condition and $\Vdash_P \dot{Q}$ has the κ-chain condition, then $P \otimes \dot{Q}$ has the κ-chain condition.

Proof. Note that since P has the κ-chain condition, we have $\Vdash_P \kappa$ is regular and uncountable, so the second hypothesis makes sense.

Suppose by way of contradiction that $A = \{(p_\alpha, \dot{q}_\alpha): \alpha < \kappa\}$ is an antichain. Let G be P-generic over V, and in V[G], let $B = \{\alpha: p_\alpha \in G\}$. If $\dot{B}$ is the natural term denoting B, then clearly $\Vdash_P \{\dot{q}_\alpha: \alpha \in \dot{B}\}$ is an antichain. Since $\Vdash_P \dot{Q}$ has the κ-chain condition, we have $\Vdash_P \exists\beta < \kappa\ \dot{B} \subseteq \beta$. Since P has the κ-chain condition, there is $\gamma < \kappa$ such that $\Vdash_P \dot{B} \subseteq \gamma$. But this contradicts the fact that if $\gamma < \alpha < \kappa$ then $p_\alpha \Vdash \alpha \in \dot{B}$. □

Remarks. 1. The converse of Theorem 2.1 is also true, i.e., if $P \otimes \dot{Q}$ has the κ-chain condition then P has the κ-chain condition and $\Vdash_P \dot{Q}$ has the κ-chain condition. We will not need this result so the proof is left to the reader.

2. It should be pointed out that if $Q \in V$ then to say that $\Vdash_P$ "Q has the κ-chain condition" is stronger than simply to say that Q has the κ-chain condition. It can happen that P and Q both have the κ-chain condition but $P \times Q$ does not. See [7] for a discussion of this topic.

Theorem 2.2. Assume (a) P_α is the direct limit of $\langle P_\beta: \beta < \alpha\rangle$

(b) if $\beta < \alpha$ then P_β has the κ-chain condition

(c) if $\mathrm{cf}\,\alpha = \kappa$ then $\{\beta < \alpha: P_\beta$ is the direct limit of $\langle P_\gamma: \gamma < \beta\rangle\}$ is stationary in α.

Then P_α has the κ-chain condition.

Proof. First note that if $p,q \in P_\alpha$, $\beta < \alpha$, support(p) $\cup$ support(q) $\subseteq \beta$ and $p|\beta$, $q|\beta$ are compatible in P_β, then p and q are compatible. (If $r \leq p|\beta, q|\beta$, define $\bar{r} \in P_\alpha$ by $\bar{r}|\beta = r$, $\bar{r}(\gamma) = \dot{1}$ for $\beta \leq \gamma < \alpha$. Then $\bar{r} \leq p,q$.)

Suppose A is an antichain in P_α of cardinality κ. Note that by (a), if $p \in P_\alpha$ then $\exists\beta < \alpha$ support(p) $\subseteq \beta$. If cf $\alpha < \kappa$ then $\exists\beta < \alpha\ \exists B \subseteq A\ |B| = \kappa$ and $\forall p \in B$ support(p) $\subseteq \beta$. But then by our remark above $\{p|\beta: p \in B\}$ would be an antichain of cardinality κ in P_β, contradicting (b).

A similar argument works if cf $\alpha > \kappa$.

Suppose cf $\alpha = \kappa$. Let $A = \{p_\xi: \xi < \kappa\}$. Let $\langle\alpha_\xi: \xi < \kappa\rangle$ be a continuous sequence of ordinals cofinal in α. Define $f: \kappa \to \kappa$ by $f(\xi)$ = the least η such that support$(p_\xi|\alpha_\xi) \subseteq \alpha_\eta$. Since $\{\beta: P_\beta$ is the direct limit of $\langle P_\gamma: \gamma < \beta\rangle\}$ is stationary in α, f is regressive on a stationary subset of κ, namely $\{\xi: \xi$ is a limit ordinal and P_{α_ξ} is the direct limit of $\langle P_\gamma: \gamma < \alpha_\xi\rangle\}$. By Fodor's Theorem, f is constant on a stationary set S. Say $f''S = \{\eta\}$. By thinning out S if necessary we may assume that if $\xi_1,\xi_2 \in S$ and $\xi_1 < \xi_2$ then support$(p_{\xi_1}) \subseteq \alpha_{\xi_2}$. Since P_{α_η} has the κ-chain condition, $\exists\xi_1,\xi_2 \in S\ \xi_1 < \xi_2$ and $p_{\xi_1}|\alpha_\eta, p_{\xi_2}|\alpha_\eta$ are compatible. Let $r \leq p_{\xi_1}|\alpha_\eta, p_{\xi_2}|\alpha_\eta$, and define $q \in P_\alpha$ by $q|\alpha_\eta = r$ and

$$q(\beta) = \begin{cases} p_{\xi_1}(\beta) & \text{if } \alpha_\eta \leq \beta < \alpha_{\xi_2} \\ p_{\xi_2}(\beta) & \text{if } \alpha_{\xi_2} \leq \beta < \alpha. \end{cases}$$

Then $q \leq p_{\xi_1}, p_{\xi_2}$, contradiction. □

Corollary 2.3. Assume that for every $\beta < \alpha$, $\Vdash_\beta \dot{Q}_\beta$ has the countable chain condition. If direct limits are taken everywhere then P_α has the countable chain condition.

Proof. By induction on α. Theorem 2.1 handles the successor case and Theorem 2.2 handles the limit case. □

Corollary 2.3 is the heart of the consistency proof for Martin's Axiom. See Section 3. Another corollary, useful in reverse Easton extensions, is the following:

Corollary 2.4. If κ is a Mahlo cardinal, $|P_\beta| < \kappa$ for all $\beta < \kappa$, and P_β is the direct limit of $\langle P_\gamma : \gamma < \beta \rangle$ whenever $\beta \le \kappa$ and β is strongly inaccessible, then P_κ has the κ-chain condition.

Now we turn to closure conditions.

Theorem 2.5. Suppose that for all $\beta < \alpha$, $\Vdash_\beta \dot{Q}_\beta$ is κ-closed. Suppose also that all limits are inverse or direct and that if $\beta \le \alpha$, β is a limit ordinal and cf $\beta < \kappa$, then P_β is the inverse limit of $\langle P_\gamma : \gamma < \beta \rangle$. Then P_α is κ-closed.

Proof. Let $\langle p_\xi \; \xi < \zeta \rangle$, $\zeta < \kappa$, be a decreasing sequence in P_α. We must find $p \le p_\xi$ for all $\xi < \zeta$. By induction on $\beta \le \alpha$, we obtain $p|\beta \in P_\beta$ such that $p|\beta \le p_\xi|\beta$ for all $\xi < \zeta$ and support $(p|\beta)$

$= \bigcup\{\text{support}(p_\xi|\beta) : \xi < \zeta\}$. Given $p|\beta$, we obtain $p(\beta)$ (hence $p|\beta+1$) as follows. Since $p|\beta \le p_\xi|\beta$ for all $\xi < \zeta$ it is clear that

$$p|\beta \Vdash \langle p_\xi(\beta) : \xi < \zeta \rangle \text{ is a decreasing sequence in } \dot{Q}_\beta .$$

If all $p_\xi(\beta) = \dot{1}$, let $p(\beta) = \dot{1}$; otherwise let $p(\beta)$ be any term such that $p|\beta \Vdash_\beta (\forall\xi<\zeta)\, p(\beta) \le p_\xi(\beta)$. This clearly works.

If β is a limit ordinal then $p|\beta$ is completely determined by the $p|\gamma$, $\gamma < \beta$, and everything is clear except possibly that $p|\beta \in P_\beta$. If support$(p|\beta)$ is not cofinal in β then this is easy, so suppose otherwise. Since $\text{support}(p|\beta) = \cup\{\text{support}(p_\xi|\beta) : \xi < \zeta\}$ we must have either $\sup(\text{support}(p_\xi|\beta)) = \beta$ for some $\xi < \zeta$ or else $\text{cf}\ \beta \le |\zeta| < \kappa$. In either case P_ξ is the inverse limit of $\langle P_\gamma : \gamma < \beta\rangle$ and $p|\beta \in P_\beta$ by virtue of the fact that $p|\gamma \in P_\gamma$ for all $\gamma < \beta$. □

<u>Corollary 2.6</u>. If P is κ-closed and $\Vdash_P \dot{Q}$ is κ-closed, then $P \otimes \dot{Q}$ is κ-closed.

A similar result holds for κ-directed closed forcing.

<u>Theorem 2.7</u>. Suppose that for all $\beta < \alpha$, $\Vdash_\beta \dot{Q}_\beta$ is κ-directed closed. Suppose also that all limits are inverse or direct and that if $\beta \le \alpha$, β is a limit ordinal and $\text{cf}\ \beta < \kappa$, then P_β is the inverse limit of $\langle P_\gamma : \gamma < \beta\rangle$. Then P_α is κ-directed closed.

The proof is left to the reader.

3. Martin's Axiom.

Here we show the relative consistency of Martin's Axiom with $2^{\aleph_0} > \aleph_1$. This was historically the first application of iterated forcing, due to Solovay and Tennenbaum [21].

Martin's Axiom (MA) is the following assertion: Let P be an arbitrary partial ordering with the countable chain condition, and let $\langle D_\alpha : \alpha < \lambda\rangle$ be a sequence of dense subsets of P with $\lambda < 2^{\aleph_0}$. Then there exists $G \subseteq P$ such that

(i) G is directed, i.e., if $p,q \in G$ then there is $r \in G$ such that $r \leq p,q$, and

(ii) $\forall\alpha < \lambda \quad G \cap D_\alpha \neq 0$.

We sometimes express (i) and (ii) by saying G is <u>generic with respect to</u> the D_α.

Note that MA is implied by CH. We will be interested in the relative consistency of MA with $\neg$ CH.

Let MA* be the same as MA except for the additional hypothesis that $|P| < 2^{\aleph_0}$.

<u>Lemma 3.1</u>. MA* implies MA.

<u>Proof</u>. Let P and $\langle D_\alpha : \alpha < \lambda\rangle$ be given with $|P|$ arbitrary. It is not difficult to see that there is $P' \subseteq P$ such that $|P'| = \lambda$ and

(*) if $p,q \in P'$ and $\exists r \in P\ r \leq p,q$ then $\exists r \in P'\ r \leq p,q$ and

(**) $D_\alpha \cap P'$ is dense in P'.

(The logicians will see that an elementary substructure of $(P, \leq, D_\alpha)_{\alpha<\lambda}$ will suffice.) By (*) compatibility in P' is the same as compatibility in P, so if P has the countable chain condition, so does P'. Now apply MA* to P' and $\langle D_\alpha \cap P' : \alpha < \lambda\rangle$ to obtain $G \subseteq P'$ satisfying (i) and (ii). But then G also works for P.

Thus it will suffice to prove the relative consistency of MA* with $\neg$ CH. Before giving the proof, however, it is necessary to have some simple facts about forcing.

<u>Lemma 3.2.</u> Suppose P has the λ-chain condition, $|P| \leq \kappa$, and $\kappa^{\mu} = \kappa$ for all $\mu < \lambda$. If $\Vdash_P |\dot{Q}| \leq \kappa$, then $|P \otimes \dot{Q}| \leq \kappa$.

<u>Proof.</u> Without loss of generality we may assume $\Vdash_P \dot{Q} \subseteq \kappa$. Fix $\dot{q}$ such that $\Vdash_P \dot{q} \in Q$. We associate with $\dot{q}$ a function f on κ such that $\forall\alpha$ $f(\alpha)$ is a maximal pairwise incompatible subset of $\{p \in P: p \Vdash \dot{q} = \alpha\}$. Note that since P has the λ-chain condition, $|\{\alpha: f(\alpha) \neq 0\}| < \lambda$. It follows that the maximum number of such functions f is κ, since $\kappa = \Sigma\{\kappa^{\mu}: \mu < \lambda\}$. But if the same function f is associated with both $\dot{q}_1$ and $\dot{q}_2$, then clearly $\Vdash_P \dot{q}_1 = \dot{q}_2$. Thus $\{\dot{q}: \Vdash_P \dot{q} \in \dot{Q}\}$ has cardinality $\leq \kappa$ and the conclusion of the lemma follows. □

<u>Lemma 3.3.</u> Suppose P has the λ-chain condition, $|P| \leq \kappa$, $\kappa^{\mu} = \kappa$ for all $\mu < \lambda$, and $\kappa^{\nu} = \kappa$. Then $\Vdash_P 2^{\nu} \leq \kappa$.

<u>Proof.</u> Fix $\dot{X}$ such that $\Vdash_P \dot{X} \subseteq \nu$. Again we associate a function f on κ such that $f(\alpha) = (A_\alpha, B_\alpha)$, where A_α is a maximal incompatible subset of $\{p \in P: p \Vdash \alpha \in \dot{X}\}$ and B_α is a maximal incompatible subset of $\{p \in P: p \Vdash \alpha \notin \dot{X}\}$. The number of such pairs (A_α, B_α) is at most $\Sigma\{\kappa^{\mu}: \mu < \lambda\} = \kappa$, so the number of such functions f is at most $\kappa^{\nu} = \kappa$. If f is associated with both $\dot{X}_1$ and $\dot{X}_2$ then clearly $\Vdash_P \dot{X}_1 = \dot{X}_2$. Thus $\Vdash_P 2^{\nu} \leq \kappa$. □

Now we are ready to attack the consistency proof.

<u>Theorem 3.4.</u> Assume GCH and let κ be a regular cardinal, $\kappa > \aleph_1$. Then there is a partial ordering P with the countable chain condition such that $\Vdash_P$ MA ans $2^{\aleph_0} = \kappa$.

<u>Proof.</u> We will obtain P as a κ-stage iteration P_κ. By induction on α we determine P_α $(\alpha \leqslant \kappa)$ and $\dot{Q}_\alpha$ $(\alpha < \kappa)$ so that $\Vdash_\alpha \dot{Q}_\alpha$ has the

countable chain condition and $|\dot{Q}_\alpha| < \kappa$.

Direct limits are always taken at limit ordinals. Hence by Corollary 2.3 each P_α has the countable chain condition. Using Lemma 3.2 we note by induction on α that $|P_\alpha| \le \kappa$. Using GCH it then follows that for all $\alpha \le \kappa$, $\Vdash_\alpha (\forall\lambda < \kappa) 2^\lambda \le \kappa$ by Lemma 3.3.

Given P_α, it remains to find $\dot{Q}_\alpha$. This will be done in such a way that for any $Q \in V[G_\kappa]$, if $|Q| < \kappa$ and Q has the countable chain condition in $V[G_\kappa]$, then Q "appears" as $\dot{Q}_\alpha$ for arbitrarily large $\alpha < \kappa$.

Fix a mapping $\pi: \kappa \to \kappa \times \kappa$ such that for every $(\beta,\gamma) \in \kappa \times \kappa$, there are arbitrarily large $\alpha < \kappa$ such that $\pi(\alpha) = (\beta,\gamma)$, and whenever $\pi(\alpha) = (\beta,\gamma)$, then $\beta \le \alpha$.

Now suppose P_α has been determined. As remarked above, $\Vdash_\alpha (\forall\lambda < \kappa) 2^\lambda \le \kappa$, so there is a term $\langle \dot{Q}^\gamma_\alpha : \gamma < \kappa \rangle$ such that

$\Vdash_\alpha$ "$\langle \dot{Q}^\gamma_\alpha : \gamma < \kappa \rangle$ enumerates all partial orderings with universe some element of κ".

Let $\pi(\alpha) = (\beta,\gamma)$. Let $\dot{Q}_\alpha$ be the term which denotes the same partial ordering Q in $V[G_\alpha]$ that $\dot{Q}^\gamma_\beta$ denotes in $V[G_\beta]$ (recall $V[G_\beta] \subseteq V[G_\alpha]$) provided Q still has the countable chain condition in $V[G_\alpha]$, and which denotes the trivial partial ordering otherwise. Note that the term $\dot{Q}_\alpha$ cannot be the same as $\dot{Q}^\gamma_\beta$ unless $\beta = \alpha$, since $\dot{Q}_\alpha$ is a term of the language of forcing over P_α and $\dot{Q}^\gamma_\beta$ is a term of the language of forcing over P_β.

Clearly $\Vdash_\alpha \dot{Q}_\alpha$ has the countable condition and $|\dot{Q}_\alpha| < \kappa$.

This completes the construction of the P_α, $\alpha \le \kappa$. For the rest of the proof we need the following lemma.

<u>Lemma 3.5</u>. If $\alpha < \kappa$, $X \subseteq \alpha$ and $X \in V[G_\kappa]$, then $(\exists\beta < \kappa)$ $X \in V[G_\beta]$.

<u>Proof</u>. Suppose X is denoted by a term $\dot{X}$ of the language of forcing. For each $\gamma < \alpha$, let A_γ be a maximal incompatible subset

of $\{p \in P_\kappa : p \Vdash \gamma \in \dot{X}\}$. Then $X = \{\gamma : G_\kappa \cap A_\gamma \neq 0\}$. Now each A_γ is countable, and since P_κ is a direct limit support(p) is bounded below κ for every $p \in A_\gamma$. Since κ is regular there is $\beta < \kappa$ such that support(p) $\subseteq \beta$ for all $p \in A_\gamma$ and all $\gamma < \alpha$. But now if $A'_\gamma = \{p|\beta : p \in A_\gamma\}$, then we have $X = \{\gamma : G_\beta \cap A'_\gamma \neq 0\}$, and $X \in V[G_\beta]$. □

Now we check MA* in $V[G_\kappa]$. Suppose $V[G_\kappa] \models$ "$|Q| < \kappa$, Q has the countable chain condition, and $\langle D_\alpha : \alpha < \lambda \rangle$ is a sequence of dense subsets of Q for some $\lambda < \kappa$". Without loss of generality we may assume $Q = (\bar{\alpha}, \leq_Q)$, where $\bar{\alpha} < \kappa$. By Lemma 3.5 there is $\beta < \kappa$ such that Q, $\langle D_\alpha : \alpha < \lambda \rangle \in V[G_\beta]$. Then Q must be the denotation of some $\dot{Q}^\gamma_\beta$ in $V[G_\beta]$. Choose α' so large that $\pi(\alpha') = (\beta,\gamma)$. Since Q has the countable chain condition in $V[G_\kappa]$ it must certainly have it in $V[G_{\alpha'}]$. Hence Q is the denotation of $\dot{Q}_{\alpha'}$. But then $V[G_{\alpha'+1}]$ contains a set G which is Q-generic over $V[G_{\alpha'}]$; hence $G \cap D_\alpha \neq 0$ for all $\alpha < \lambda$ and G is directed.

It remains only to check that $\Vdash_\kappa 2^{\aleph_0} = \kappa$. By Lemma 3.3 we have $\Vdash_\kappa 2^{\aleph_0} \leq \kappa$. By the argument in the preceding paragraph, if Q is the ordering for adding one Cohen real, then Q is the denotation of $\dot{Q}_\alpha$ for arbitrarily large $\alpha < \kappa$; each time we force with $\dot{Q}_\alpha$ one more real is added. Hence $\Vdash_\kappa 2^{\aleph_0} \geq \kappa$. This completes the proof. □

<u>Remark</u>. GCH was needed only in order to find a regular cardinal $\kappa > \aleph_1$ such that $2^\lambda \leq \kappa$ for all $\lambda < \kappa$. Any such κ would work.

4. Generalized Martin's Axiom.

The search for extensions of MA for larger cardinals has proved to be rather difficult. One natural candidate for such an extension is the following: If P is countably closed and has the $\aleph_2$-chain condition and $\langle D_\alpha : \alpha < \lambda\rangle$, $\lambda < 2^{\aleph_1}$, is a sequence of dense subsets of P, then there is $G \subseteq P$ generic with respect to the D_α. Unfortunately, it is not yet known whether this assertion is relatively consistent (as one would hope) with CH + $2^{\aleph_1} > \aleph_2$.

On the other hand, several weaker versions are known to be consistent. The one described here is perhaps the simplest (and weakest), due to the author. The first generalization of MA is an unpublished result of Laver. The strongest version is due to Shelah [15]. Herink [8] has strengthed the version which appears here.

A partial ordering P is $\aleph_1$-linked iff there exists $f: P \to \omega_1$ such that $(\forall\alpha \in \omega_1) f^{-1}\{\alpha\}$ is pairwise compatible. We refer to f as a linking function for P. Note that if P is $\aleph_1$-linked then P has the $\aleph_2$-chain condition.

Let us call P countably compact iff for any countable set $A \subseteq P$, if for every finite $F \subseteq A$ there is $p \in P$ such that $(\forall q \in F)\ p \le q$, then there is $p \in P$ such that $(\forall q \in A)\ p \le q$. Note that if P is countably compact, then P is $\aleph_1$-directed closed and hence $\aleph_1$-closed.

The generalized version of MA which we will consider is the following:

GMA: If P is countably compact and $\aleph_1$-linked and $\langle D_\alpha : \alpha < \lambda\rangle$, $\lambda < 2^{\aleph_1}$, is a sequence of dense subsets of P, then there is $G \subseteq P$ such that G is generic with respect to the D_α.

The consistency proof for GMA is quite similar to the one for MA, so we shall concentrate on the important differences. There are two: First, the obvious version of GMA* does not seem to imply GMA in every possible case, and second, the fact that the iteration P_κ has the $\aleph_2$-chain is no longer a trivial consequence of Theorems 2.1 and 2.2.

Let GMA* denote GMA restricted to orderings P such that $|P| < 2^{\aleph_1}$. The analogue of Lemma 3.1 is

Lemma 4.1. Assume that $(\forall\lambda < 2^{\aleph_1})\ \lambda^{\aleph_0} < 2^{\aleph_1}$. Then GMA* implies GMA.

Proof. Suppose P is countably compact and $\aleph_1$-linked, and $\langle D_\alpha: \alpha < \lambda\rangle$ is a sequence of dense subsets of P. Construct by induction on $\xi < \omega_1$ an increasing sequence $\langle Q_\xi: \xi < \omega_1\rangle$ of subsets of P such that

(i) $|Q_\xi| \le \lambda^{\aleph_0}$

(ii) $(\forall\alpha < \lambda)(\forall p \in Q_\xi)(\exists q \in D_\alpha \cap Q_{\xi+1})\ q \le p$

(iii) if ξ is limit and A is a countable (or finite) subset of $\cup\{Q_\eta: \eta < \xi\}$ such that $(\exists p \in P)(\forall q \in A)\ p \le q$, then $(\exists p \in Q_\xi)(\forall q \in A)\ p \le q$.

Then $Q = \cup\{Q_\xi: \xi < \omega_1\}$ is countably compact and $\aleph_1$-linked and $(\forall\alpha < \lambda)\ D_\alpha \cap Q$ is dense in Q. Now apply GMA* to Q. □

Theorem 4.2. Assume GCH, and let $\kappa > \aleph_2$ be a regular cardinal. Then there is a countably closed partial ordering P with the $\aleph_2$-chain condition such that $\Vdash_P$ GMA and $2^{\aleph_0} = \aleph_1$ and $2^{\aleph_1} = \kappa$.

Proof. As before, P will be a κ-stage iteration P_κ. For all $\alpha < \kappa$ we will have $\Vdash_\alpha$ "$\dot{Q}_\alpha$ is countably compact and $\aleph_1$-linked and $|\dot{Q}_\alpha| \le \kappa$". If α is a limit ordinal and cf $\alpha = \omega$ then P_α is the inverse limit of the $P_\beta, \beta < \alpha$; if cf$\alpha > \omega$ then P_α is the direct limit of the $P_\beta, \beta < \alpha$. By Theorem 2.5 P_κ is $\aleph_1$-closed. Note that support(p) is countable or finite for each $p \in P_\kappa$.

We assert that P_κ has the $\aleph_2$-chain condition. This almost follows from Theorems 2.1 and 2.2 by induction on $\alpha \le \kappa$; only the case cf$\alpha = \omega$ is not covered. It turns out, however, that the proof for this case is no easier than the proof for the general case, so we give a direct proof for P_κ.

For each $\alpha < \kappa$ let $\dot{f}_\alpha$ be a term such that

$\Vdash_\alpha \dot{f}_\alpha$ is a linking function for $\dot{Q}_\alpha$.

Now fix $p \in P_\kappa$. If $\alpha < \kappa$ then there must be $r \in P_\alpha$ such that $r \le p|\alpha$ and for some $\xi < \omega_1$, $r \Vdash_\alpha \dot{f}_\alpha(p(\alpha)) = \xi$. Define $q \in P_\kappa$ so that $q|\alpha = r$ and $q(\beta) = p(\beta)$ for $\alpha \le \beta < \kappa$. Then $q \le p$ and $q|\alpha \Vdash_\alpha \dot{f}_\alpha(p(\alpha)) = \xi$.

Using this remark repeatedly it is easy to find $\langle p_n : n \in \omega\rangle$, $\langle \alpha_n : n \in \omega\rangle$ and $\langle \xi_n : n \in \omega\rangle$ such that

(1) $\forall n\ p_{n+1} \le p_n \le p$.

(2) $\forall n\ p_{n+1}|\alpha_n \Vdash_{\alpha_n} \dot{f}_{\alpha_n}(p_n(\alpha_n)) = \xi_n$

(3) $\forall \alpha \in \cup\{\text{support}(p_n): n \in \omega\}$, $\{n: \alpha_n = \alpha\}$ is infinite.

Let $\sigma(p) = \langle p_n : n \in \omega\rangle$, let $A(p) = \{\alpha_n : n \in \omega\}$, and define g_p on ω by $g_p(n) = (\alpha_n, \xi_n)$.

Suppose I is a pairwise incompatible subset of P_κ and $|I| = \aleph_2$. A standard Δ-system argument (using CH) yields $I' \subseteq I$, $|I'| = \aleph_2$ such that $\{A(p): p \in I'\}$ forms a Δ-system with kernel D (i.e., $A(p) \cap A(q) = D$ whenever $p,q \in I'$ and $p \neq q$). Moreover, we may assume that there is a function h such that for all $p \in I'$, $g_p|\{n: \alpha_n \in D\} = h$ (Since in any case there are only $\aleph_1$ such functions $g_p|\{n: \alpha_n \in D\}$, assuming CH).

We assert that any two elements of I' are compatible; this contradiction will establish the $\aleph_2$-chain condition. Let $p,q \in I'$, $\sigma(p) = \langle p_n : n \in \omega\rangle$, $\sigma(q) = \langle q_n : n \in \omega\rangle$. We must find $r \le p,q$. We will determine $r|\alpha$ by induction on $\alpha < \kappa$ so that $\forall n\ r|\alpha \le p_n|\alpha, q_n|\alpha$ and if $\forall n\ p_n(\beta) = q_n(\beta) = \dot{1}$, then $r(\beta) = \dot{1}$. The limit case is trivial. Supppose $\alpha = \beta+1$. If $p_n(\beta) = q_n(\beta) = \dot{1}$, let $r(\beta) = \dot{1}$. If not, then $\beta \in A(p) \cup A(q)$. Suppose $\beta \in A(p) - A(q)$. (The case $\beta \in A(q) - A(p)$ is symmetric.) Since

$r|\beta \leq p_n|\beta$ for all n, we clearly have

$$r|\beta \Vdash_\beta \langle p_n(\beta) : n \in \omega\rangle \text{ is a decreasing sequence in } \dot{Q}_\beta.$$

Since $\Vdash_\beta \dot{Q}_\beta$ is countably closed, there is $r(\beta)$ such that $r|\beta \Vdash_\beta \forall n(r(\beta) \leq p_n(\beta))$. Then $r|\beta+1 \leq p_n|\beta+1, q_n|\beta+1$ for all n.

Finally, suppose $\beta \in A(p) \cap A(q) = D$. As above,

$$(*)\quad r|\beta \Vdash_\beta \langle p_n(\beta) : n \in \omega\rangle, \langle q_n(\beta) : n \in \omega\rangle \text{ are decreasing.}$$

We claim that for each n,

$$(**)\quad r|\beta \Vdash_\beta p_n(\beta), q_n(\beta) \text{ are compatible.}$$

Fix $m \geq n$ such that $h(m) = (\beta,\xi)$. Then $r|\beta \Vdash \dot{f}_\beta(p_m(\beta)) = \dot{f}_\beta(q_m(\beta)) = \xi$, so $r|\beta \Vdash_\beta p_m(\beta), q_m(\beta)$ are compatible, and (**) follows by (*). But (*) and (**) together with the assumption that $\Vdash_\beta$ "$\dot{Q}_\beta$ is countably compact" imply that for some $r(\beta)$,

$$r|\beta \Vdash_\beta \forall n\, r(\beta) \leq p_n(\beta), q_n(\beta).$$

This completes the proof of the $\aleph_2$-chain condition.

If now $(\forall\lambda < \kappa)\ \lambda^{\aleph_0} < \kappa$ then we may finish the proof using Lemma 3.1 just as in the case of ordinary MA. If $(\exists\lambda < \kappa)\ \lambda^{\aleph_0} = \kappa$ then we must be circumspect.

Let Q be a partial ordering. A finite or countable set $A \subseteq Q$ is <u>consistent</u> iff for any finite $F \subseteq A$ there is $p \in Q$ such that for all $q \in F$, $p \leq q$. Let $\bar{Q}$ be the set of all consistent subsets of Q, partially ordered by reverse inclusion (i.e., $A \leq B$ iff $B \subseteq A$). It is easy to see that $\bar{Q}$ is countably compact.

We claim that if Q is $\aleph_1$-linked, then so is $\bar{Q}$. Let $f: Q \to \omega_1$ be a linking function. Let us say that $A, B \in \bar{Q}$ are <u>isomorphic</u> if there is a bijection $\phi: A \to B$ such that

(i) for $p, q \in A$, $p \leq q$ iff $\phi(p) \leq \phi(q)$, and

(ii) for all $p \in A$, $f(p) = f(\phi(p))$.

It is clear by CH that there are only $\aleph_1$ isomorphism types. If $g(A)$ = the isomorphism type of A, then we assert that g is (essentially) a linking function for $\bar{Q}$. This requires showing that if A and B are isomorphic (via ϕ, say) then A and B are compatible, i.e., $A \cup B$ is consistent. Suppose $F \subseteq A \cup B$ is finite. Find finite $F_1 \subseteq A$ and $F_2 \subseteq B$ such that ϕ carries F_1 onto F_2 and $F \subseteq F_1 \cup F_2$. Fix p such that $p \leq q$ for all $q \in F_1$. Then $f(p) = f(\phi(p))$ so there is $r \in Q$ such that $r \leq p, \phi(p)$. But then $r \leq q$ for all $q \in F_1 \cup F_2$.

Now at stage α in the iteration we determine $\dot{Q}_\alpha$ as follows. Let $\pi(\alpha) = (\beta,\gamma)$, and let Q be the denotation of $\dot{Q}^\gamma_\beta$ in $V[G_\beta]$. Let $\dot{Q}_\alpha$ be the term which denotes $\bar{Q}$ if Q is $\aleph_1$-linked (in $V[G_\alpha]$), and which denotes the trivial ordering otherwise.

We must verify that this works. In $V[G_\kappa]$, suppose Q is countably compact and $\aleph_1$-linked, and $\langle D_\alpha : \alpha < \lambda \rangle$, $\lambda < \kappa$ is a sequence of dense subsets of Q. Let $I_\alpha \subseteq D_\alpha$ be maximal incompatible. Then $|I_\alpha| \leq \aleph_2$. As in the previous section, find $Q_1 \subseteq Q$ so that $|Q_1| = \lambda$, $(\forall\alpha < \lambda) I_\alpha \subseteq Q_1$ and if $p,q \in Q_1$ and $(\exists r \in Q) r \leq p,q$, then $(\exists r \in Q_1) r \leq p,q$. Then Q_1 is still $\aleph_1$-linked, so there is a linking function $f: Q_1 \to \omega_1$.

As in the previous section, there is $\beta > \kappa$ such that $Q_1, f, I_\alpha \in V[G_\beta]$ for all $\alpha < \lambda$. We may assume Q_1 is the denotation of $\dot{Q}^\gamma_\beta$ in $V[G_\beta]$. If $\pi(\alpha) = (\beta,\gamma)$, then clearly the denotation of $\dot{Q}_\alpha$ is $\bar{Q}_1$.

Let $E_\alpha = \{A \in \bar{Q}_1 : I_\alpha \cap A \neq 0\}$. We claim E_α is dense in $\bar{Q}_1$ (and clearly belongs to $V[G_\alpha]$). If $A \in \bar{Q}_1$, then since Q is countably compact there is $p \in Q$ with $p \leq q$ for all $q \in A$. Since I_α is maximal in Q there is $p' \in I_\alpha$ such that p and p' are compatible. But then $\{p'\} \cup A \leq A$ and $\{p'\} \cup A \in E_\alpha$.

Now $V[G_{\alpha+1}]$ contains a set G which is $\bar{Q}_1$-generic over $V[G_\alpha]$. Let $H = \cup G$. Since $G \cap E_\alpha \neq 0$, $H \cap I_\alpha \neq 0$. Also every finite subset of H is compatible. We need only check that H is directed. But without loss of generality we may assume that for all $\alpha,\beta < \lambda$ there is $\gamma < \lambda$ such that $D_\gamma = \{p \in Q: (\exists q \in I_\alpha)(\exists \in I_\beta) p \leq q,r\}$. Then $H \cap \cup\{I_\alpha : \alpha < \lambda\}$ is directed and is generic with respect to the D_α. □

Remark. As in the last section, GCH was not necessary. All we needed to assume was CH, κ is regular, and $(\forall\lambda < \kappa) 2^\lambda \leq \kappa$.

5. Intermediate Stages.

It seems clear intuitively that if P_α is an α-stage iteration and $\beta < \alpha$, then forcing with P_α should be the same as forcing with P_β followed by an $(\alpha-\beta)$-stage iteration in the sense of $V[G_\beta]$. Here we make this intuition precise.

If $\beta < \alpha$ and $p \in P_\alpha$, let $p^\beta = p|\{\gamma: \beta \le \gamma < \alpha\}$. Thus $p = (p|\beta) \cup p^\beta$. Let $P_{\beta\alpha} = \{p^\beta: p \in P_\alpha.\}$ Given a P_β-generic set G_β, define an ordering on $P_{\beta\alpha}$ by setting $f \le g$ iff $(\exists p \in G_\beta) p \cup f \le p \cup g$. (The last $\le$ is in P_α. It is an exercise to verify transitivity.) Let $\dot{P}_{\beta\alpha}$ be a term of the language of forcing with P_β which denotes $P_{\beta\alpha}$ with the ordering above.

Theorem 5.1. P_α is isomorphic to a dense subset of $P_\beta \otimes \dot{P}_{\beta\alpha}$.

Proof. Let $\phi(p) = (p|\beta, p^\beta)$. If $p \le q$ then $p|\beta \le q|\beta$ and it follows that $p \le (p|\beta) \cup q^\beta$. Hence $p|\beta \Vdash_\beta p^\beta \le q^\beta$, and $\phi(p) \le \phi(q)$. It is also easy to show that if $\phi(p) \le \phi(q)$ then $p \le q$. Suppose $(p,\dot{f}) \in P_\beta \otimes \dot{P}_{\beta\alpha}$. Then there is $q \le p$ and $f \in P_{\beta\alpha}$ such that $q \Vdash \dot{f} = f$. But then $\phi(q \cup f) = (q,f) \le (p,\dot{f})$, and the range of ϕ is dense in $P_\beta \otimes \dot{P}_{\beta\alpha}$. □

Theorem 5.2. Let α and β be as above. Then

$$\Vdash_\beta \dot{P}_{\beta\alpha} \text{ is isomorphic to an } (\alpha-\beta)\text{-stage iteration.}$$

Proof. We work in $V' = V[G_\beta]$. By induction on $\gamma \le \alpha - \beta$ we will construct a γ-stage iteration $\underset{\sim}{P}_\gamma$ (with associated terms $\dot{\underset{\sim}{Q}}_\gamma$) and an isomorphism $h_\gamma: P_{\beta,\beta+\gamma} \to \underset{\sim}{P}_\gamma$ such that

(*) If $\delta < \gamma$ and $f \in P_{\beta,\beta+\gamma}$ then $(h_\gamma(f))|\delta = h_\delta(f|\beta+\delta)$.

Note that (*) allows us to handle limit ordinals easily; simply define $\underset{\sim}{P}_\gamma$ to be all γ-sequences s such that for some $f \in P_{\beta,\beta+\gamma}$, $\forall\delta<\gamma$ $h_\delta(f|\beta+\delta) = s|\delta$. Details in this case are left to the reader.

Now suppose $\underset{\sim}{P}_\gamma$ has been obtained. We must find $\dot{\underset{\sim}{Q}}_\gamma$ and

$h_{\gamma+1}$. The natural choice for $\dot{\underline{Q}}_\gamma$ is $\dot{Q}_{\beta+\gamma}$, but this is technically incorrect because $\dot{Q}_{\beta+\gamma}$ is a term in the language of forcing over V with $P_{\beta+\gamma}$, whereas $\dot{\underline{Q}}_\gamma$ must be a term of the language of forcing over V' with $\underline{P}_\gamma$. Nevertheless it can be arranged that $\dot{\underline{Q}}_\gamma$ and $\dot{Q}_{\beta+\gamma}$ both denote the same object. This is done as follows.

Suppose H is $\underline{P}_\gamma$-generic over V'. Then $h_\gamma^{-1}(H)$ is $P_{\beta,\beta+\gamma}$-generic over V'. As in Theorem 1.1(a), $G_\beta \otimes h_\gamma^{-1}(H)$ is $P_\beta \otimes \dot{P}_{\beta,\beta+\gamma}$-generic over V, and if ϕ is as in Theorem 5.1, then $\phi^{-1}(G_\beta \otimes h_\gamma^{-1}(H))$ is $P_{\beta+\gamma}$-generic over V. Moreover, $V'[H] = V[\phi^{-1}(G_\beta \otimes h_\gamma^{-1}(H))]$. Now let $\dot{\underline{Q}}_\gamma$ be the canonical term representing the same object in V'[H] that is represented by $\dot{Q}_{\beta+\gamma}$ in $V[\phi^{-1}(G_\beta \otimes h_\gamma^{-1}(H))]$.

Since $\Vdash_{\beta+\gamma} \dot{Q}_{\beta+\gamma}$ is a partial ordering, we must have also $\Vdash_{\underline{P}_\gamma} \dot{\underline{Q}}_\gamma$ is a partial ordering; this determines $\underline{P}_{\gamma+1}$.

If $\Vdash_{\beta+\gamma} \tau \in \dot{Q}_{\beta+\gamma}$, then just as before there is a canonical term τ' which represents in V'[H] the same object that τ represents in $V[\phi^{-1}(G_\beta \otimes f_\gamma^{-1}(H))]$. Define $h_{\gamma+1}: P_{\beta,\beta+\gamma+1} \to \underline{P}_{\gamma+1}$ by $h_{\gamma+1}(f) = p$ iff $p|\gamma = h_\gamma(f|\beta+\gamma)$ and $p(\gamma) = f(\beta+\gamma)'$. It is easy to see that $h_{\gamma+1}$ is an isomorphism of $P_{\beta,\beta+\gamma+1}$ into $\underline{P}_{\gamma+1}$.

We must show $h_{\gamma+1}$ is onto. It will suffice to show that if σ is a term such that $\Vdash_{\underline{P}_\gamma} \sigma \in \dot{\underline{Q}}_\gamma$, then there is τ such that $\Vdash_{\beta+\gamma} \tau \in \dot{Q}_{\beta+\gamma}$, and $\Vdash_{\underline{P}_\gamma} \sigma = \tau'$. Now $\underline{P}_\gamma$, $\dot{\underline{Q}}_\gamma$ and σ are all elements of $V' = V[G_\beta]$. Since $\underline{P}_\gamma$ and $\dot{\underline{Q}}_\gamma$ are canonically determined from G_β, there are canonical terms $\dot{P}$ and $\dot{Q}$ of the language of forcing over V with P_β which represent them. (Note that $\dot{Q}$ is a term denoting a <u>term</u>.) It is not difficult to find a term $\dot{\sigma}$ representing σ in $V[G_\beta]$ such that

$$\Vdash_\beta \text{"}\dot{\sigma} \text{ and } \dot{Q} \text{ are terms and } \Vdash_{\dot{P}} \dot{\sigma} \in \dot{Q}.\text{"}$$

If $G_{\beta+\gamma}$ is $P_{\beta+\gamma}$-generic over V, then $\phi(G_{\beta+\gamma})$ determines a generic set $G_\beta \otimes G^\beta \subseteq P_\beta \otimes \dot{P}_{\beta\alpha}$. If $H = h_\gamma''(G^\beta)$, then H is $\underline{P}_\gamma$-generic over $V[G_\beta]$. Now let τ be the term of the language of forcing over V with $P_{\beta+\gamma}$ obtained as follows. Given $G_{\beta+\gamma}$, compute $\dot{\sigma}^{V[G_\beta]}$ and then compute the denotation s of this term in $V[G_\beta][H]$. Let τ denote s.

Naturally τ makes use of canonical terms denoting G_β and H, but such terms are easily obtained from the term $\dot{G}_{\beta+\gamma}$ denoting $G_{\beta+\gamma}$. It is easy to see that τ works, and this completes the proof. □

Remarks 1. In a sense Theorem 5.2 is completely trivial, and it should always be thought of as such. The subtleties about terms of different languages may be confusing at first, but they must be brought to the reader's attention. Since the correspondences are invariably canonical, however, henceforth we shall treat, say, $\dot{\underline{Q}}_\gamma$ and $\dot{Q}_{\beta+\gamma}$ as being virtually the same thing.

2. In view of Theorem 5.2, we may speak of $P_{\beta\alpha}$ as being the inverse (or direct) limit of $P_{\beta\gamma}, \gamma < \alpha$, even though technically $P_{\beta\alpha}$ is not an iteration.

One question which will come up later on is the following: If P_α is the inverse limit of the $P_\gamma, \gamma < \alpha$, does it follow that in $V[G_\beta]$, $P_{\beta\alpha}$ is the inverse limit of the $P_{\beta\gamma}, \gamma < \alpha$? The answer in general is no, although in practice the answer is usually yes. We formulate a condition in Theorem 5.4 below which will make the affirmative answer easy to verify in most cases. First, however, we treat the case of direct limits.

Theorem 5.3. If P_α is the direct limit of the $P_\gamma, \gamma < \alpha$, then in $V[G_\beta]$, $P_{\beta\alpha}$ is the direct limit of the $P_{\beta\gamma}, \gamma < \alpha$.

Proof. Simply observe that every member of $P_{\beta\alpha}$ ends in a string of $\dot{1}$'s. □

As before, suppose P_α is an α-stage iteration and $\beta < \alpha$. Let $K_\beta = \{\delta \le \alpha : \beta < \delta$ and P_δ is the direct limit of the $P_\gamma, \gamma < \delta\}$. Call a set $X \subseteq \alpha - \beta$ K_β-thin iff $(\forall\delta \in K_\beta)\sup(X\cap\delta) < \delta$. Note that if $p \in P_{\beta\alpha}$ then support(p) is K_β-thin.

Theorem 5.4. Suppose that for every limit ordinal $\gamma \le \alpha$, P_γ is either the direct or inverse limit of the $P_\delta, \delta < \gamma$. Fix $\beta < \alpha$, and assume that $V[G_\beta] \models$ For every K_β-thin set X there is K_β-thin $Y \in V$ such that $X \subseteq Y$.

If P_α is the inverse limit of the $P_\gamma, \gamma < \alpha$, then in $V[G_\beta]$, $P_{\beta\alpha}$ is the inverse limit of the $P_{\beta\gamma}, \gamma < \alpha$.

Proof. Suppose $f \in V[G_\beta]$ is such that dom $f = \{\gamma : \beta \le \gamma < \alpha\}$ and $\forall\gamma(\beta \le \gamma < \alpha \to f|\gamma \in P_{\beta\gamma})$. We must find $g \in P_{\beta\alpha}$ so that $f \le g \le f$. (This means that for all γ such that $\beta \le \gamma < \alpha$, $f|\gamma \le g|\gamma \le f|\gamma$.)

Let $\dot{f}$ be a term denoting f in $V[G_\beta]$, and such that $\Vdash_\beta \dot{f} \in P_{\beta\alpha}$. Suppose $Y \in V$ is K_β-thin, $p_0 \Vdash_\beta$ support$(\dot{f}) \subseteq Y$, and $p_0 \in G_\beta$. Now for $\beta \le \gamma < \alpha$, if $\gamma \in Y$ let $g(\gamma)$ be the term of the language of forcing with P_γ which denotes the object x obtained from G_γ as follows: First compute the denotation of $\dot{f}$ in $V[G_\beta]$. This is possible since $\gamma \ge \beta$. Then $\dot{f}^{V[G_\beta]^\beta}(\gamma)$ is a term of the language of forcing with P_γ which in $V[G_\gamma]$ denotes a set x. Then $g(\gamma)$ denotes x. If $\beta \le \gamma < \alpha$ and $\gamma \notin Y$ let $g(\gamma) = \dot{1}$. In either case $\Vdash_\gamma g(\gamma) \in \dot{Q}_\gamma$.

We claim that $p \cup g \in P_\alpha$, where p is a β-sequence of $\dot{1}$'s. It is clear that if $q = p \cup g$, then $\forall\gamma \Vdash_\gamma q(\gamma) \in \dot{Q}_\gamma$ so the only difficulty occurs at limits. If P_δ is the direct limit of $P_\gamma, \gamma < \delta$, then support(q) $\cap\ \delta$ is bounded in δ since support(q) = Y, which is K_β-thin. Since the only other alternative is an inverse limit we must have $p \cup g \in P_\alpha$, and thus $g \in P_{\beta\alpha}$. But clearly $p_0 \Vdash_\beta \dot{f} \le g \le \dot{f}$. □

A good example of the use of Theorem 5.4 is given by the iteration P_κ of Section 4. If $\beta < \kappa$ then the K_β-thin sets are precisely the countable (or finite) subsets of $\kappa - \beta$. But since P_κ is countably closed, these sets are the same in V and in $V[G_\kappa]$. Thus the hypothesis of Theorem 5.4 is satisfied. It follows that from the viewpoint of any intermediate stage $V[G_\beta]$, the rest of the extension (via $\underset{\to}{P}_\kappa$) looks very much like the entire original extension (via P_κ). This accords very well with intuition.

Another corollary of Theorem 5.4 is the following, which will be used in Section 6.

Theorem 5.5. Suppose κ is a regular uncountable cardinal, $\beta < \alpha$, P_β has the κ-chain condition and if $\beta \leq \gamma < \alpha$ then $\Vdash_\gamma \dot{Q}_\gamma$ is κ-directed closed. Suppose also that if $\beta < \gamma \leq \alpha$ and γ is limit then P_γ is the direct or inverse limit of the P_δ, $\delta < \gamma$, and if $\mathrm{cf}\gamma < \kappa$ then P_γ is the inverse limit of the P_δ, $\delta < \gamma$. Then

$$\Vdash_\beta \dot{P}_{\beta\alpha} \text{ is } \kappa\text{-directed closed.}$$

Proof. By Theorems 2.7, 5.3 and 5.4, it will suffice to show that

$$\Vdash_\beta \text{"}\forall X \text{ if } X \text{ is } K_\beta\text{-thin then } (\exists Y \in V)\ X \subseteq Y \text{ and } Y \text{ is } K_\beta\text{-thin"}$$

Suppose $p \Vdash_\beta \dot{X}$ is K_β-thin. Let $Y = \{\xi : (\exists q \leq p)\, q \Vdash_\beta \xi \in \dot{X}\}$. Obviously $p \Vdash_\beta \dot{X} \subseteq Y$. We claim Y is K_β-thin. If $\delta \in K_\beta$ then $\mathrm{cf}\delta \geq \kappa$. Now $p \Vdash_\beta \sup(\dot{X} \cap \delta) < \delta$, and since P_β has the κ-chain condition there must be $\delta' < \delta$ such that $p \Vdash_\beta \sup(\dot{X} \cap \delta) \leq \delta'$. But then $Y \cap \delta \subseteq \delta'$ and we're done. □

6. Reverse Easton Forcing.

If κ is a regular cardinal and $\lambda \geq \kappa$, then let $P(\kappa,\lambda)$ be the usual ordering for adding λ subsets of κ. Elements of $P(\kappa,\lambda)$ are functions p such that $|p| \leq \kappa$, $\mathrm{dom}(p) \subseteq \lambda$ and range $(p) \subseteq 2$. Let $p \leq q$ iff $p \supseteq q$.

Easton [6] showed that a model of $2^{\aleph_0} = \aleph_2 \wedge 2^{\aleph_1} = \aleph_3$ results if one begins with a model of GCH and forces with the product $P(\aleph_0,\aleph_2) \times P(\aleph_1,\aleph_3)$. This amounts to adding the subsets of ω_1 <u>before</u> the subsets of ω are added. This observation remains true even when subsets of many cardinals are added at once as in [6]. One can always regard the subsets of larger cardinals as having been added before the subsets of smaller cardinals.

What happens if, as may seem more natural, this process is reversed? In the example above, if we force first with $P(\aleph_0,\aleph_2)$ and then force over that model with $P(\aleph_1,\aleph_3)$ <u>as defined in that model</u>, then $\aleph_2$ will be collapsed to $\aleph_1$ and we will arrive at a model of CH! Nevertheless it still turns out that this <u>reverse</u> Easton forcing has many applications. As Silver [20] has shown, it is extremely useful in a number of problems involving large cardinals. One of the simplest of these problems is treated in this section.

A treatment of reverse Easton forcing from a Boolean-algebraic point of view may be found in [13].

For the rest of this section we will assume that the reader is familiar with the elementary theory of measurable cardinals. See [9].

A cardinal κ is λ-<u>supercompact</u> iff there is a transitive class M and an elementary embedding $j: V \to M$ such that

(a) j is the identity on V_κ, the set of sets of rank $< \kappa$

(b) if $x \subseteq M$ and $|x| \leq \lambda$ then $x \in M$.

(c) $j(\kappa) > \kappa$.

The following is a slightly weakened version of a result due to Silver. See the remarks at the end of the section.

Theorem 6.1. Suppose κ is κ^{++}-supercompact and $2^{\kappa} = \kappa^{+}$. Then there is a partial ordering P such that

$$\Vdash_{P} \kappa \text{ is measurable and } 2^{\kappa} = \kappa^{++}.$$

Proof. We will obtain P as a $(\kappa+1)$stage iteration. Given P_{α}, obtain $\dot{Q}_{\alpha}$ as follows. Let $\dot{\mu}_{\alpha}$ be a term such that

$$\Vdash_{\alpha} \mu_{\alpha} \text{ is the least regular cardinal } \geq \aleph_{\beta+3}^{V} \text{ for all } \beta < \alpha.$$

Then let $\dot{Q}_{\alpha}$ be $P(\dot{\mu}_{\alpha}, \dot{\mu}_{\alpha}^{++})$, i.e.,

$$\Vdash_{\alpha} \dot{Q}_{\alpha} = P(\dot{\mu}_{\alpha}\ \dot{\mu}_{\alpha}^{++}).$$

If α is a limit ordinal but α is not strongly inaccessible, let P_{α} be the inverse limit of the $P_{\beta}, \beta < \alpha$; if α is strongly inaccessible let P_{α} be the direct limit of the $P_{\beta}, \beta < \alpha$.

Let $P = P_{\kappa+1}$. If G is P-generic, then we assert that in V[G], κ is measurable and $2^{\kappa} = \kappa^{++}$. We need several lemmas.

Lemma 6.2. P_{κ} has the κ-chain condition, $|P_{\kappa}| = \kappa$ and $\Vdash_{\kappa} \dot{\mu}_{\kappa} = \kappa$.

Proof. Since κ is κ^{++}-supercompact, κ is measurable, hence Mahlo, hence strongly inaccessible. In view of this fact, it is easy to check by induction that if $\alpha < \kappa$ then $|P_{\alpha}| < \kappa$ and $\Vdash_{\alpha} \dot{\mu}_{\alpha} < \kappa$. (Use Lemma 3.2 for the successor case.) Since P_{κ} is the direct limit of the P_{α}, $\alpha < \kappa$, it follows immediately that $|P_{\kappa}| = \kappa$. Also, since κ is Mahlo Corollary 2.4 applies to show that P_{κ} has the κ-chain condition. Thus $\Vdash_{\kappa} \kappa$ is regular. This implies immediately $\Vdash_{\kappa} \dot{\mu}_{\kappa} = \kappa$. □

Since $|P_{\alpha}| < \kappa$ for all $\alpha < \kappa$ we shall assume that

$(\forall\alpha < \kappa)P_\alpha \in V_\kappa$, and hence $j(P_\alpha) = P_\alpha$ and j is the identity on P_α. (Recall $j: V \to M$ by κ^{++}-supercompactness). Of course this assumption depends on the way the forcing apparatus has been set up; if it should turn out not to be the case that $P_\alpha \in V_\kappa$ it is still true that j yields a canonical isomorphism of P_α and $j(P_\alpha)$.

Now in M, $j(P)$ is an iteration of length $j(\kappa+1) = j(\kappa) + 1$. We shall denote the stages of this iteration by P^M_α. The assumption above then says $P_\alpha = P^M_\alpha$ for all $\alpha < \kappa$. Also, since κ remains strongly inaccessible in M, we have $P_\kappa = P^M_\kappa$.

<u>Lemma 6.3</u>. $P_{\kappa+1} = P^M_{\kappa+1}$.

<u>Proof</u>. Suppose $\Vdash_\kappa \dot{q} \in \dot{Q}_\kappa$. Since P_κ has the κ-chain condition, κ^{++} is the same whether computed in V or $V[G_\kappa]$. Thus $\Vdash_\kappa$ "domain $(\dot{q}) \subseteq \kappa^{++}$ and $|\text{domain}(\dot{q})| < \kappa$". Again by the κ-chain condition, there must be $D \subseteq \kappa^{++}$, $|D| < \kappa$, such that $\Vdash_\kappa$ domain $(\dot{q}) \subseteq D$. For each $\alpha \in D$ let A_α be a maximal incompatible subset of $\{p \in P_\kappa: p \Vdash_\kappa \dot{q}(\alpha) = 0\}$, and let B_α be a maximal incompatible subset of $\{p \in P_\kappa: p \Vdash \dot{q}(\alpha) = 1\}$. Note that if $\dot{q}_1$ and $\dot{q}_2$ both give rise to the same sequences $\langle A_\alpha: \alpha \in D\rangle$ and $\langle B_\alpha: \alpha \in D\rangle$ then $\Vdash_\kappa \dot{q}_1 = \dot{q}_2$. Thus $\langle A_\alpha: \alpha \in D\rangle$ and $\langle B_\alpha: \alpha \in D\rangle$ completely describe $\dot{q}$. Moreover, by condition (b) of the definition of κ^{++}-supercompactness, M and V both have the same sequences $\langle A_\alpha: \alpha \in D\rangle$ and $\langle B_\alpha: \alpha \in D\rangle$. Thus $P_{\kappa+1}$ will come out the same whether it is defined in M or in V. □

Now G is $P_{\kappa+1}$-generic over V; by Lemma 6.3 G is also $P^M_{\kappa+1}$-generic over M.

<u>Lemma 6.4</u>. If $x \subseteq M[G]$, $x \in V[G]$ and $M[G] \models |x| \leq \kappa^{++}$, then $x \in M[G]$.

Proof. Since M[G] is a model of AC, it will suffice to assume that x is a set of ordinals. Let $\dot{x}$ be a term denoting x in V[G]. By the κ-chain condition for P_κ, there is a set $D \in V$ such that $|D| \leq \kappa^{++}$ and $\Vdash_\kappa \dot{x} \subseteq D$. For $\alpha \in D$ let A_α be a maximal incompatible subset of $\{p \in P_\kappa: p \Vdash_\kappa \alpha \in \dot{x}\}$. By (b) in the definition of κ^{++}-supercompact, $\langle A_\alpha: \alpha \in D\rangle \in M$. But then $x \in M[G]$ since $x = \{\alpha \in D: A_\alpha \cap G \neq 0\}$. □

For notational convenience let us denote $\kappa + 1$ henceforth by ξ. The ordering on $P^M_{\xi,j(\xi)}$ (recall that in $M, P^M_{j(\xi)} = P^M_\xi \otimes P^M_{\xi,j(\xi)}$) is definable in M[G], hence in V[G].

Lemma 6.5. $P^M_{\xi,j(\xi)}$ is κ^{+++}-directed closed in V[G].

Proof. If $A \subseteq P^M_{\xi,j(\xi)}$, every finite subset of A has a lower bound in A, and $|A| \leq \kappa^{++}$, then $A \in M[G]$ by Lemma 6.4. Now in M[G], $P^M_{\xi,j(\xi)}$ is κ^{+++}-directed closed by Theorem 5.5 so A has a lower bound. □

Now let $A = \{q^\xi: (\exists p \in G) q = j(p)\}$.

Lemma 6.6. Every finite subset of A has a lower bound in A.

Proof. Suppose $p_i \in G$ and $q_i = j(p_i)$, $i = 1,\dots,n$. Then there is $p \in G$ such that $p \leq p_i$, $i = 1,\dots,n$. Let $q = j(p)$. Since P_κ is a direct limit, there is some $C \subseteq \kappa$, $\sup C < \kappa$, such that $\text{support}(p) \subseteq C \cup \{\kappa\}$. Since j is elementary, $\text{support}(q) \subseteq j(C) \cup \{j(\kappa)\} = C \cup \{j(\kappa)\}$. Moreover, we have $q|\kappa = p|\kappa$. Since $p \in M$ we have $p \cup q^\xi \in P^M_{j(\xi)}$. Also, since $q(\kappa) = \dot{1}$, $p \cup q^\xi \leq q$. But $q \leq q_i$ so $p \cup q^\xi \leq p \cup q^\xi_i$, $i = 1,\dots,n$. And by the definition of $P^M_{\xi,j(\xi)}$, this means $q^\xi \leq q^\xi_i$, $i = 1,\dots,n$.

It follows now from Lemma 6.5 that A has a lower bound $q_0 \in P^M_{\xi,j(\xi)}$. Let H be $P^M_{\xi,j(\xi)}$-generic over V[G] with $q_0 = H$. Let $V_1 = V[G]$ and $M_1 = M[G][H]$.

<u>Lemma 6.7.</u> In $V_1[H]$ the map j can be extended to an elementary embedding $j^* : V_1 \to M_1$.

<u>Proof.</u> Let $G' = G \otimes H$. If $p \in G$ and $q = j(p)$ then $p \leq q|\kappa + 1$, and $q^{\xi} \in H$ since $q_0 \in H$ and $q_0 \leq q^{\xi}$. Therefore, $q \in G'$. So $p \in G$ implies $j(p) \in G'$.

Define j^* as follows. If $x \in V_1$ then there is some term $\dot{x}$ such that $x = \dot{x}^{V[G]}$. Let $j^*(x) = j(\dot{x})^{M[G']}$. We claim j^* is elementary. Suppose $V_1 \models \phi(x_1,\ldots,x_n)$. Choosing terms $\dot{x}_1,\ldots,\dot{x}_n$ as above, there must be $p \in G$ such that $p \Vdash_{\xi} \phi(\dot{x}_1,\ldots,\dot{x}_n)$. Since j is elementary, $j(p) \Vdash^{M}_{j(\xi)} \phi(j(\dot{x}_1),\ldots,j(\dot{x}_n))$. But $j(p) \in G'$, so $M_1 \models \phi(j^*(x_1),\ldots,j^*(x_n))$. Note that this shows j^* is well-defined, since if $p \Vdash_{\xi} \dot{x}_1 = \dot{x}_2$ then $j(p) \Vdash^{M}_{j(\xi)} j(\dot{x}_1) = j(\dot{x}_2)$. □

Now we are nearly done. Let $U = \{X \subseteq \kappa\colon X \in V_1$ and $\kappa \in j^*(X)\}$. By Lemma 6.5, every subset of κ in $V_1[H]$ is already in V_1, so U is a measure ultrafilter on κ in $V_1[H]$. It is easy to see that $\Vdash_{\kappa+1} 2^{\kappa} = \kappa^{++}$ ($\Vdash_{\kappa} 2^{\kappa} = \kappa^{+}$ and $\dot{Q}_{\kappa}$ is $P(\kappa,\kappa^{++})$) so $V_1[H] \models |U| \leq \kappa^{++}$, and by Lemma 6.5 again, $U \in V_1$. Therefore $\Vdash_{\kappa+1}$ "$2^{\kappa} = \kappa^{+}$ and κ is measurable", and the proof of Theorem 6.1 is complete. □

<u>Remarks</u> 1. In order to obtain the consistency of the assertion, "There is a measurable cardinal κ such that $2^{\kappa} = \kappa^{++}$" it is not sufficient to assume simply the consistency of the existence of a measurable cardinal. Kunen [10] has shown that if κ is measurable and $2^{\kappa} \geq \kappa^{++}$ then there are inner models with many measurable cardinals.

2. The argument we have given is not quite optimal, since by means of a trick one can show that in the situation we considered κ is actually κ^{+}-supercompact in $V[G]$. The same trick would yield our conclusion that κ is measurable in $V[G]$ from the weaker assumption that κ is κ^{+}-supercompact in V. Since

the trick tends to obscure the important ideas, however, we have omitted it.

7. Axiom A forcing.

In this section we consider a type of iteration which covers a great number of cases arising in practice.

A partial ordering $(P,\leq)$ satisfies Axiom A iff there exist partial orderings $<\leq_n : n \in \omega>$ such that

(1) $p \leq_0 q$ iff $p \leq q$

(2) if $p \leq_{n+1} q$ then $p \leq_n q$

(3) if $<p_n : n \in \omega>$ is such that $p_{n+1} \leq_n p_n$ for all n, then $(\exists q \in P)(\forall n)\ q \leq_n p_n$.

(4) if I is a pairwise incompatible subset of P, then $(\forall p \in P)(\forall n)(\exists q \in P)\ q \leq_n p$ and $\{r \in I: q$ is compatible with $r\}$ is countable.

Condition (4) may be rephrased in terms of forcing as follows:

(4') $(\forall p \in P)(\forall n)$ if $p \Vdash \dot{a} \in V$, then there is a countable set $x \in V$ and $q \leq_n p$ such that $q \Vdash \dot{a} \in x$.

To see that (4) implies (4') suppose $p \Vdash \dot{a} \in V$. Let I be a maximal incompatible subset of $\{r \in P: \exists a_r \in V\ q \Vdash \dot{a} = a_r\}$. If $q \leq_n p$ and $\{r \in I: q$ and r are compatible$\}$ is countable, then $q \Vdash \dot{a} \in x$, where $x = \{a_r : r \in I$ and q and r are compatible$\}$.

For the other direction, let I be a pairwise incompatible set. Without loss of generality we may assume I is maximal. Let $\dot{a}$ denote the unique element of I belonging to the P-generic set G. Given $p \in P$ we must have $p \Vdash \dot{a} \in V$. By (4') there is $q \leq_n p$ and countable x such that $q \Vdash \dot{a} \in x$. But then $\{r \in I: q,r$ are compatible$\} \subseteq x$.

Following are some examples. In each case details are left to the reader.

Example 1. Countable chain condition forcing. If P has the countable chain condition, then for $n \geq 1$, let $p \leq_n q$ iff $p = q$.

2. Countably closed forcing. Let $p \leq_n q$ iff $p \leq q$.

3. Perfect-set forcing (Sacks forcing). Conditions are sets $p \subseteq \cup\{^n2\colon n \in \omega\}$ satisfying

(a) $\forall s \in p\ \forall n \in \text{domain}(s)\ s|n \in p$.

(b) $\forall s \in p\ \ \exists t,u \in p\ \ s \subseteq t,u$ and t and u are incomparable with respect to inclusion.

Let $p \le q$ iff $p \subseteq q$. Conditions may be thought of as trees of height ω in which forking occurs above each element. Such trees are called <u>perfect</u> since $\{f \in {}^{\omega}2\colon (\forall n) f|n \in p\}$ is then a perfect subset of ${}^{\omega}2$ (with the product topology). If p is perfect and $s \in p$ then the <u>degree</u> <u>of</u> s <u>in</u> p is $|\{n\colon \exists t \in p\ s|n = t|n \text{ and } s(n) \neq t(n)\}|$. This is just the number of f ks below s in the tree p. Let $p \le_n q$ iff $p \le q$ and $(\forall s \in p)$ if the degree of s in q is $\le_n$, then $s \in p$. See [1] for a treatment of iterated perfect-set forcing.

4. Prikry-Silver forcing. Conditions are functions p such that domain(p) $\subseteq \omega$, range(p) $\subseteq 2$ and ω-domain(p) is infinite. Let $p \le q$ iff $p \supseteq q$. Let $p \le_n q$ iff $p \supseteq q$ and the first n members of ω-domain(p) and ω-domain(q) are the same.

5. Laver forcing. Conditions are sets $p \subseteq \cup\{^n\omega\colon n \in \omega\}$ such that

(a) $(\forall s \in p)(\forall n \in \text{domain}(s))\ s|n \in p$.

(b) $(\exists s \in p)(\forall t \in p)\ t \subseteq s$ or $s \subseteq t$, and if $s \subseteq t$ then $\{u \in p\colon u$ is an immediate $\subseteq$-successor of $t\}$ is infinite.

Let $p \le q$ iff $p \subseteq q$. The definition of $\le_n$ is left as an exercise for the reader. See [11].

6. Mathias forcing. Conditions are pairs (s,A), where s is a finite subset of ω and A is an infinite subset of ω. Let $(s,A) \le (t,B)$ iff $s \supseteq t$, $A \subseteq B$ and $s - t \subseteq B$. This ordering will be discussed at length in Section 9.

7. Adding a closed unbounded set with finite conditions. This is perhaps the simplest cardinal-preserving ordering which does not satisfy Axiom A. Conditions are finite functions p

from ω_1 into ω_1 such that for some closed unbounded set $C \subseteq \omega_1$, if $f: \omega_1 \to \omega_1$ enumerates C in increasing order, then $p \subseteq f$. Let $p \leq q$ iff $p \supseteq q$.

In work at present unpublished, Shelah [17] has found a class of partial orderings, the proper partial orderings, which is more extensive than the class of Axiom A orderings, and for which the analogue of Theorem 7.1 below is still true. For example, the ordering in Example 7 is proper but fails to satisfy Axiom A. Nevertheless we have decided to treat only the Axiom A case, for two reasons. First, the Axiom A orderings are nearly as extensive as the proper orderings, and iterated Axiom A forcing seems to be a little easier to handle than iterated proper forcing. Second, in the author's opinion the proof of Theorem 7.1 is conceptually simpler than the proof for proper forcing. The reader who understands the proof of 7.1 should be able to follow Shelah's argument without great difficulty.

We shall be interested here in iteration with countable support, i.e. iterations in which inverse limits are taken at ordinals of cofinality ω and direct limits are taken elsewhere. For example, the iteration in Section 4 has countable support. One checks easily that for any element p of such an iteration, support(p) is countable or finite.

Theorem 7.1. Suppose $\mathbb{P}_\alpha$ is an α-stage iteration with countable support, and $(\forall\beta < \alpha) \Vdash_\beta \dot{Q}_\beta$ satisfies Axiom A. Then $\mathbb{P}_\alpha$ does not collapse ω_1. In addition, if $\alpha \leq \omega_2$, CH holds, and $(\forall\beta < \alpha) \Vdash_\beta |\dot{Q}_\beta| \leq \aleph_1$, then $\mathbb{P}_\alpha$ has the $\aleph_2$-chain condition.

Proof. The fundamental notion in treating Axiom A forcing is the following. Suppose $F \subseteq \gamma \leq \alpha$, F is finite, and $n \in \omega$. If $p, q \in P_\gamma$ then let $p \leq_{F,n} q$ iff $p \leq q$ and $(\forall\beta \in F)\ p|\beta \Vdash_\beta p(\beta) \leq_n q(\beta)$. (Here, of course, $\leq_n$ refers to the ordering on $\dot{Q}_\beta$.) A set $D \subseteq P_\gamma$ is (F,n)-dense iff $(\forall p \in P_\gamma)(\exists q \in D)\ q \leq_{F,n} p$. Note that $\leq_{F,n}$ is a partial ordering.

A sequence $\langle (p_n, F_n): n \in \omega \rangle$ will be called a fusion sequence iff

(a) $\forall n\ p_{n+1} \leq_{F_n,n} p_n$

(b) $\forall n\ F_n \subseteq F_{n+1}$

(c) $\bigcup\{F_n : n \in \omega\} = \bigcup\{\text{support}(p_n) : n \in \omega\}$.

<u>Lemma 7.2.</u> If $<(p_n,F_n) : n \in \omega>$ is a fusion sequence in P_α, then there is a $p \in P_\alpha$ such that $\forall n\ p \leq_{F_n,n} p_n$.

<u>Proof.</u> By induction on $\beta < \alpha$ we determine $p|\beta \in P_\beta$ so that $\forall n\ p|\beta \leq_{F_n \cap \beta, n} p_n|\beta$, and if $p_n(\beta) = \dot{1}$ for all n, then $p(\beta) = \dot{1}$. The latter condition guarantees that support $(p|\beta)$ is at most countable, and it makes the case β limit a triviality. Suppose $B = \gamma+1$. If $\gamma \notin \bigcup\{\text{support } p_n : n \in \omega\}$ then let $p(\gamma) = \dot{1}$. Otherwise fix n minimal such that $\gamma \in F_n$. Let $\dot{q}_m = p_m(\gamma)$ if $m \geq n$ and let $\dot{q}_m = p_n(\gamma)$ if $m < n$. Now $p|\gamma \leq p_n|\gamma$ for all n, so $p|\gamma \Vdash (\forall m)\ \dot{q}_{m+1} \leq_m \dot{q}_m$. Hence there is $p(\gamma)$ such that $p|\gamma \Vdash (\forall m) p(\gamma) \leq_m \dot{q}_m$.

<u>Lemma 7.3.</u> Suppose $\beta \leq \alpha$. Then:

(a) If $\Vdash_\beta \dot{a} \in V$, then $\{q \in P_\beta : \exists$ countable $x \in V$, $q \Vdash_\beta \dot{a} \in x\}$ is (F,n)-dense for all finite $F \subseteq \beta$ and $n \in \omega$.

(b) If $\Vdash_\beta$ "$\dot{x}$ is a countable subset of V", then $\{q \in P_\beta :$ $\exists$ countable $x \in V$, $q \Vdash_\beta \dot{x} \subseteq x\}$ is (F,n)-dense for all finite $F \subseteq \beta$ and $n \in \omega$.

(c) If $\beta < \gamma \leq \alpha$ and $\Vdash \dot{f} \in P_{\beta\gamma}$, then $\{q \in P_\beta :$ $(\exists f \in P_{\beta\gamma})\ q \Vdash_\beta \dot{f} = f\}$ is (F,n)-dense for all finite $F \subseteq \beta$ and all $n \in \omega$.

Proof. The proof is by induction on β

(a) First suppose $\beta = \gamma + 1$. If $p \Vdash_\beta \dot{a} \in V$, then $p|\gamma \Vdash_\gamma$ "$p(\gamma) \Vdash_{\dot{Q}_\gamma} \dot{a} \in V$". (Note that $\dot{a}$ is here being used in two different senses. See the remarks in Section 5.) Hence by condition (4') applied to $\dot{Q}_\gamma$, there must be terms $\dot{x}$ and $\dot{q}$ of the language of forcing with P_γ such that

$$p|\gamma \Vdash_\gamma \text{"}\dot{x} \text{ is a countable subset of } V,$$
$$\dot{q} \le_n p(\gamma), \text{ and } \dot{q} \Vdash_{\dot{Q}_\gamma} \dot{a} \in \dot{x}\text{"}$$

Now, using (b) of the inductive hypothesis, there is countable $x \in V$ and $q' \le_{F\cap\gamma,n} p|\gamma$ such that $q' \Vdash_\gamma \dot{x} \subseteq x$. Define $q \in P_\gamma$ by $q|\gamma = q'$ and $q(\gamma) = \dot{q}$. Clearly $q \Vdash_\beta \dot{a} \in x$, and $q \le_{F,n} p$.

Now suppose β is limit. Fix F and n, and suppose γ is such that $F \subseteq \gamma < \beta$. If $p \in P_\beta$, let p^γ denote $p|\{\delta: \gamma \le \delta < \beta\}$. Then $p|\gamma \Vdash_\gamma$ "$p^\gamma \Vdash \dot{a} \in V$", where the last "$\Vdash$" refers to forcing with $P_{\gamma\beta}$. Hence there must be $\dot{f}$ and $\dot{b}$ such that

$$p|\gamma \Vdash_\gamma \text{"}\dot{f} \in P_{\gamma\beta},\ \dot{f} \le p^\gamma,\ \dot{b} \in V \text{ and } \dot{f} \Vdash \dot{a} = \dot{b}\text{"}.$$

(Here we are using $\dot{a}$ both as a term of the language of forcing with P_β and as the corresponding term of the language of forcing over $V[G_\gamma]$ with $P_{\gamma\beta}$.) By (a) and (c) of the inductive hypothesis applied to P_γ there must be $q \le_{F,n} p|\gamma$, $f \in P_{\gamma\beta}$ and countable $x \in V$ such that $q \Vdash_\gamma \dot{f} = f$ and $\dot{b} \in x$. But then $q \cup f \in P_\beta$, $q \cup f \le_{F,n} p$, and $q \cup f \Vdash_\beta \dot{a} \in X$. This proves (a).

(b) Suppose $p \Vdash_\beta \dot{x} = \langle \dot{a}_n : n \in \omega \rangle$. Fix $F \subseteq \beta$ and n. Using (a) repeatedly, it is easy to construct a fusion sequence $\langle (p_m, F_m) : m \in \omega \rangle$ such that $p_m = p$ for $m < n$, $F_n = F$, and for $m = n + i$ there is countable $x_i \in V$ such that $p_m \Vdash_\beta \dot{a}_i \in x_i$. Let $x = \cup\{x_i : i \in \omega\}$. If $q \in P$ is such that $q \le_{F_m,m} p_m$ for all m, then $q \le_{F,n} p$ and $q \Vdash_\beta \dot{x} \subseteq x$.

(c) Suppose $p \in P_\beta$ and $\Vdash_\beta \dot{f} \in \dot{P}_{\beta\gamma}$. By (b) there is $q \leq_{F,n} p$ and countable $x \in V$ such that $q \Vdash_\beta \text{support}(\dot{f}) \subseteq x$. Now define $f \in P_{\beta\gamma}$ as follows. If $\delta \notin x$ then $f(\delta) = \dot{1}$. If $\delta \in x$ then $f(\delta)$ is the term denoting the same object denoted by $g(\delta)$, where g is the interpretation of the term $\dot{f}$. It should be clear that $f \in P_{\beta\gamma}$. But also $q \Vdash_\beta \dot{f} = f$. □

Lemma 2.3(b) shows that ω_1 cannot be collapsed by forcing with P_α. There is also a version of Lemma 2.3(a) which is related to 2.3(a) as (4) is related to (4'), namely:

(a') If $I \subseteq P_\alpha$ is pairwise incompatible, then $\{q \in P_\alpha : \{p \in I : p,q \text{ are compatible}\} \text{ is countable}\}$ is (F,n)-dense for all finite $F \subseteq \alpha$ and all $n \in \omega$.

Now we prove the final sentence of Theorem 7.1. It suffices to prove that P_α has the $\aleph_2$-chain condition for all $\alpha < \omega_2$; the case for $\alpha = \omega_2$ will then follow by Theorem 2.2. The following lemma will therefore complete the proof.

<u>Lemma 7.4</u>. Assume CH. If $\alpha < \omega_2$ and $(\forall \beta < \alpha) \Vdash_\beta |\dot{Q}_\beta| \leq \aleph_1$, then there is $D_\alpha \subseteq P_\alpha$ such that $|D_\alpha| \leq \aleph_1$ and D_α is (F,n)-dense for all finite $F \subseteq \alpha$ and all $n \in \omega$.

<u>Proof</u>. The proof is by induction on α.

Suppose $\alpha = \beta + 1$. Let $D_\beta \subseteq P_\beta$ be as in the Lemma. Fix $\langle \dot{d}_\xi : \xi < \omega_1 \rangle$ such that $\Vdash_\beta \langle \dot{d}_\xi : \xi < \omega_1 \rangle$ enumerates $\dot{Q}_\beta$. For each countable pairwise incompatible set $I \subseteq D_\beta$ and each function $f: I \to \omega_1$, let $\dot{q} = \dot{q}(f)$ be the term determined as follows: if $I \cap G_\beta = \{p\}$ then $\dot{q}$ denotes the same thing as $\dot{d}_{f(p)}$; if $I \cap G_\beta = 0$ then $\dot{q}$ denotes the same thing as $\dot{1}$. Now by CH the total number of such functions f is at most $\aleph_1$ so the set T of all such terms $\dot{q}(f)$ has cardinality $\leq \aleph_1$.

Let $D_\alpha = \{p \in P_\alpha : p|\beta \in D_\beta \text{ and } p(\beta) \in T\}$. Then $|D_\alpha| \leq \aleph_1$ and we claim D_α works. Fix $p \in P_\alpha$, F and n. Let I be a maximal incompatible subset of $\{d \in D_\beta$: $d \leq p$ and for some ξ, $d \Vdash_\beta p(\beta) = \dot{d}_\xi\}$. By (a') above we may find $d \in D_\beta$ such that $d \leq_{F,n} p$ and $J = \{q \in I$: d,q are compatible$\}$ is countable. Define $f\colon J \to \omega_1$ by $f(q) = \xi$ iff $q \Vdash_\beta p(\beta) = \dot{d}_\xi$. Now it is easy to see that $d \Vdash_\beta p(\beta) = \dot{q}(f)$. Hence if $r \in D_\alpha$ is such that $r|\beta = d$ and $r(\beta) = \dot{q}(f)$, we have $r \leq p$.

Finally, suppose α is a limit ordinal. If $\beta < \alpha$ and $q \in D_\beta$ then let $\bar{q} \in P_\alpha$ be such that $\bar{q}|\beta = q$ and $\bar{q}(\gamma) = \dot{1}$ for $\beta \leq \gamma < \alpha$. Using CH, it is easy to find a set $D_\alpha \subseteq P_\alpha$ such that $|D_\alpha| \leq \aleph_1$ and

(i) if $\beta < \alpha$ and $q \in D_\beta$ then $\bar{q} \in D_\alpha$

(ii) if $<(p_n,F): n \in \omega>$ is a fusion sequence and $\forall n\ p_n \in D_\alpha$ then there is $p \in D_\alpha$ such that $\forall n\ p \leq_{F_n,n} p_n$.

We claim D_α works. Fix $p \in P_\alpha$, F and n.

Case 1. $\mathrm{cf}(\alpha) > \omega$. Then there is $\beta < \alpha$ such that support(p) $\subseteq \beta$ and $F \subseteq \beta$. By inductive hypothesis there is $q \in D_\beta$ such that $q \leq_{F,n} p|\beta$. But then $\bar{q} \leq_{F,n} p$.

Case 2. $\mathrm{cf}\ \alpha = \omega$. Let $<\alpha_m: m \in \omega>$ be an increasing sequence with supremum α, and suppose $F \subseteq \alpha_0$. For each $m \geq n$, let $f_m \in P_{\alpha_m,\alpha_{m+1}}$ be defined by $f_m(\delta) = p(\delta)$ for $\alpha_m \leq \delta < \alpha_{m+1}$. Let $q \in D_{\alpha_n}$ be such that $q \leq_{F,n} p|\alpha_n$. Now it is easy to find a fusion sequence $<(p_m,F_m): m \in \omega>$ such that $p_m = \bar{q}$ for all $m \leq n$, $F_n = F$, and for $m \geq n$, $p_{m+1} = \bar{q}_{m+1}$, where q_{m+1} is such that $q_{m+1} \in D_{\alpha_{m+1}}$ and $q_{m+1} \leq_{F_m,m} q_m \cup f_m$.

By (ii) there is $q \in D_\alpha$ such that $q \leq_{F_m,m} p_m$ for all m. But it is easy to check that $q \leq_{F,m} p$. □ □

8. An application to trees.

The rest of this paper is devoted to applications of the ideas in Section 7. In this section we will show how, given an inaccessible cardinal, to construct a partial ordering P such that

$$\Vdash_P \text{MA} + \text{"Every } \aleph_1\text{-tree has at most } \aleph_1 \text{ uncountable branches"} + 2^{\aleph_0} = \aleph_2.$$

A partial ordering $(T,\le)$ is a tree iff $(\forall t \in T)\{s \in T: s < t\}$ is well-ordered by $\le$. The level of $t \in T$ is the order type of $\{s \in T: s < t\}$. The αth level of T is $\{t \in T$: the level of t is $\alpha\}$. The height of T is the least α such that the αth level of T is empty. A branch through T is a maximal linearly ordered subset.

An Aronszajn tree is a tree T of height ω_1 with no uncountable branches such that for all $\alpha < \omega_1$ the αth level of T is countable. A Souslin tree is an Aronszajn tree in which every pairwise incomparable set is countable. An $\aleph_1$-tree is a tree of height ω_1 and cardinality $\aleph_1$. A Canadian tree is an $\aleph_1$-tree with at least $\aleph_2$ uncountable branches. A Kurepa tree is a Canadian tree in which the αth level is countable for all $\alpha \in \omega_1$.

An $\aleph_1$-tree T is special iff there is a function $f: T \to \omega$ such that $\forall s,t,u \in T$ if $s \le t$, u and $f(s) = f(t) = f(u)$, then t and u are comparable. [Note that this is different from the usual definition, which says T is special iff $(\exists f: T \to \omega)$ $(\forall s,t \in T)$ if $f(s) = f(t)$ then s and t are incomparable. It is easy to see that a tree is special in the usual sense iff it is special in our sense and it has no uncountable branches. Thus our definition extends the usual one to trees with branches.]

Theorem 8.1. If T is special then there are at most $\aleph_1$ uncountable branches through T.

Proof. Let $f: T \to \omega$ be as in the definition, and let B be a branch. There must be $s \in B$ such that $\{t \in B: f(t) = f(s)\}$ is uncountable. But then B is determined by s since $B = \{t \in T: t \le s \text{ or else } \exists u\ s \le u \text{ and } f(u) = f(s)\}$. Since there are only $\aleph_1$ choices for s, there are only $\aleph_1$ choices for B.

Theorem 8.2. Suppose $(T,\le)$ is an $\aleph_1$-tree with at most $\aleph_1$ uncountable branches. Then there is a partial ordering P with the countable chain condition such that $\Vdash_P$ is special.

Proof. Without loss of generality, we may assume T has exactly $\aleph_1$ uncountable branches $\langle B_\alpha : \alpha < \omega_1 \rangle$ (otherwise simply enlarge T). For each $\alpha < \omega_1$, let s_α be the $\leq$-minimal element of $B_\alpha - \cup\{B_\beta : \beta < \alpha\}$. Let $T' = \{t: \exists\alpha\ s_\alpha < t \in B_\alpha\}$, and let $S = T - T'$. Now S, considered as a substructure of T, is still a tree. Moreover S has no uncountable branches: if B were such a branch then for some α, $B \subseteq B_\alpha$, and then clearly $B \cap T' \neq 0$, a contradiction.

Suppose there were $f: S \to \omega$ such that if $f(s) = f(t)$ then s and t are incomparable. Then f could be extended to all of T by defining it on T' so that $f(t) = f(s_\alpha)$ if $s_\alpha < t \in B_\alpha$. But now it is easy to see that f shows T is special.

Thus we need only find P such that for some $\dot{f}$,

$$\Vdash_P \dot{f}: S \to \omega \text{ is as above.}$$

Let P consist of all finite functions p mapping S into ω such that if $p(s) = p(t)$ then s and t are incomparable. Let $p \leq q$ iff $p \supseteq q$. A proof that P has the countable chain condition may be constructed along the lines of [2], but for the reader's convenience we outline a (slightly different) proof here.

Suppose $I = \{p_\alpha : \alpha \in \omega_1\}$ is a pairwise incompatible set. Without loss of generality we may assume $|p_\alpha| = n$ for all $\alpha \in \omega_1$, and that n is minimal; also that domain $(p_\alpha) \cap \text{domain}(p_\beta) = 0$ when $\alpha \neq \beta$. Also, by thinning out I if necessary, we may assume that whenever $\alpha < \beta$, $p_\alpha(s) = p_\beta(t)$, $s \neq t$ and s and t are comparable (which must happen for some s and t if p_α and p_β are incompatible) then $s < t$. Let U be a uniform ultrafilter on ω_1, and for each $\alpha \in \omega_1$ let $\text{domain}(p_\alpha) = \{s_0^\alpha, \ldots, s_{n-1}^\alpha\}$. Now for each α there must be $i(\alpha)$, $j(\alpha) < n$ such that $\{\beta \in \omega_1 : s_{i(\alpha)}^\alpha < t_{j(\alpha)}^\beta\} \in U$ by the assumption above. Furthermore there must be i and j such that $A = \{\alpha : i(\alpha) = i \text{ and } j(\alpha) = j\} \in U$. But now if $\alpha_1, \alpha_2 \in A$ then there must be $\beta > \alpha_1, \alpha_2$ such that $s_i^\alpha < t_j^\beta$ for $\alpha = \alpha_1, \alpha_2$.

But since S is a tree, that means $s_i^{\alpha_1}$ and $s_i^{\alpha_2}$ are comparable. Hence $\{s_i^\alpha: \alpha \in A\}$ may be extended to an uncountable branch through S, contradiction. □

From the proof may be extracted the following:

Corollary 8.3. If every $\aleph_1$-tree with no uncountable branches is special, then every $\aleph_1$-tree with at most $\aleph_1$ uncountable branches is special.

Corollary 8.4. MA + $2^{\aleph_0} > \aleph_1$ implies that every $\aleph_1$-tree with at most $\aleph_1$ uncountable branches is special.

Two more technical theorems will prepare us for the main result.

A partial ordering P has property K iff every uncountable subset of P contains an uncountable pairwise compatible set. Property K clearly implies the countable chain condition. As is well-known, the usual ordering $P(\aleph_0,\lambda)$ for adding λ Cohen subsets of ω (see Section 6) has property K.

Theorem 8.5. Suppose T is an $\aleph_1$-tree and P has property K. Then forcing with P adds no new uncountable branches through T.

Proof. By way of contradiction, suppose $p \Vdash \dot{B}$ is a new branch. Let $S = \{s \in T: (\exists q \leq p)\ q \Vdash s \in \dot{B}\}$. It is obvious that for each $\alpha \in \omega_1$ there is $s_\alpha \in S$ and $p_\alpha \leq p$ such that $p_\alpha \Vdash s_\alpha \in \dot{B}$ and s_α has level α in T. Since P has property K there is uncountable $A \subseteq \omega_1$ such that $\{p_\alpha: \alpha \in A\}$ is pairwise compatible. But then $\{s_\alpha: \alpha \in A\}$ is linearly ordered, so there is an uncountable branch B through S. Since $p \Vdash \dot{B} \notin V$, it follows that $(\forall s \in S)(\exists t, u \in S)\ s \leq t,u$ and t and u are incomparable. Hence $S' = \{s \in S: s \text{ is } \leq\text{-minimal such that } s \notin B\}$ is an uncountable pairwise incomparable set. But if s and t are incomparable, $q_1 \Vdash s \in \dot{B}$, and $q_2 \Vdash t \in \dot{B}$, then q_1 and q_2 are incompatible.

Since P has the countable chain condition S can therefore have no uncountable pairwise incomparable set, contradiction.

Theorem 8.6. Suppose T is an $\aleph_1$-tree, $2^{\aleph_0} > \aleph_1$ and P is countably closed. Then forcing with P adds no new branches through T.

Proof. As before, suppose $p \Vdash \dot{B}$ is a new branch. For each $\sigma \in \cup\{{}^n2: n \in \omega\}$ we will find $p_\sigma \leq p$ and s_σ such that $p_\sigma \Vdash s_\sigma \in \dot{B}$. Also, if $\sigma \subseteq \tau$ then $s_\sigma \leq s_\tau$ and if σ and τ are incomparable (with respect to $\subseteq$) then s_σ and s_τ are incomparable. We proceed by induction on $|\sigma|$. If $\sigma = 0$ let $p_0 \leq p$ and s_0 be arbitrary such that $p_0 \Vdash s_0 \in \dot{B}$. Given p_σ and s_σ, let τ_1 and τ_2 be such that $|\tau_1| = |\tau_2| = |\sigma| + 1$ and $\sigma = \tau_1 \cap \tau_2$ (i.e., τ_1 and τ_2 are the two immediate successors of σ). Since $p_\sigma \Vdash \dot{B}$ is a new branch, there exist incomparable s_{τ_1} and $s_{\tau_2} > s_\sigma$ and there exist $p_{\tau_1}, p_{\tau_2} \leq p_\sigma$ so that $p_{\tau_i} \Vdash s_{\tau_i} \in \dot{B}$, $i = 1,2$. This completes the construction.

For $f \in {}^\omega 2$ let $p_f \leq p_{f|n}$ for all $n \in \omega$. This is possible since P is countably closed. We may also assume that p_f is chosen so that for some $s_f > s_{f|n}$, $n \in \omega$, $p_f \Vdash s_f \in \dot{B}$. But if $f \neq g$ then s_f and s_g are incomparable. Since $2^{\aleph_0} > \aleph_1$ this means $|T| > \aleph_1$, contradiction.

Corollary 8.7. If T is an $\aleph_1$-tree then there is a partial ordering P(T) satisfying Axiom A such that

$$\Vdash_{P(T)} T \text{ is special.}$$

Proof. Let $P_1 = P(\aleph_0, \aleph_2)$, the ordering for adding $\aleph_2$ Cohen reals with finite conditions. Let $\dot{P}_2$ be such that $\Vdash_{P_1}$ "$\dot{P}_2$ is the standard partial ordering for collapsing $2^{\aleph_1}$ onto $\aleph_1$ with countable conditions." (The standard ordering for collapsing λ onto $\aleph_1$ consists of countable functions mapping ω_1 into λ.) By Theorems 8.5 and 8.6, forcing with $P_1 \otimes \dot{P}_2$ adds no new branches through T, while at the same time,

$\Vdash_{P_1 \otimes \dot{P}_2}$ T has at most $\aleph_1$ uncountable branches.

Let $\dot{P}_3$ be such that $\Vdash_{P_1 \otimes \dot{P}_2}$ "$\dot{P}_3$ is the partial ordering with the countable chain condition which makes T special".

Let $P(T) = (P_1 \otimes \dot{P}_2) \otimes \dot{P}_3$. Clearly $\Vdash_{P(T)}$ T is special. To see that P(T) satisfies Axiom A it will suffice to observe that if P satisfies Axiom A and $\Vdash_P \dot{Q}$ satisfies Axiom A, then $P \otimes \dot{Q}$ satisfies Axiom A. Simply let $(p_1,\dot{q}_1) \leq_n (p_2,\dot{q}_2)$ iff $p_1 \leq_n p_2$ and $p_1 \Vdash_P \dot{q}_1 \leq_n \dot{q}_2$. Details are left to the reader. □

Now we are ready to prove the main result.

Theorem 8.8. If there is a strongly inaccessible cardinal, then there is a partial ordering P such that

$$\Vdash_P \text{MA} + 2^{\aleph_0} = \aleph_2 + \text{"Every } \aleph_1\text{-tree is special".}$$

Proof. The proof is very similar to Theorem 3.4. Let κ be strongly inaccessible. We will construct P as a κ-stage iteration P_κ with countable support (recall this means inverse limits at ordinals with cofinality ω; direct limits elsewhere). For each $\alpha < \kappa$ we will have

$\Vdash_\alpha \dot{Q}_\alpha$ satisfies Axiom A and $|\dot{Q}_\alpha| < \kappa$.

Using the inaccessibility of κ it is easy to check by induction on $\alpha < \kappa$ that $|P_\alpha| < \kappa$ (use Lemma 3.2 for successor ordinals). Hence each P_α has the κ-chain condition so by Theorem 2.2, P_κ has the κ-chain condition.

It remains to determine $\dot{Q}_\alpha$. As in Section 3, for each $\beta < \kappa$, let $\langle \dot{Q}^\gamma_\beta : \gamma < \kappa \rangle$ be such that

$\Vdash_\beta$ "$\langle \dot{Q}^\gamma_\beta : \gamma < \kappa \rangle$ enumerates all partial orderings with universe an element of κ".

Also, let $\langle \dot{T}^\gamma_\beta : \gamma < \dot{\lambda}_\beta \rangle$ be such that

$\Vdash_\beta$ "$\langle \dot{T}^\gamma_\beta : \gamma < \dot{\lambda}_\beta \rangle$ enumerates all $\aleph_1$-trees with universe ω_1".

Here $\dot{\lambda}_\beta$ is a term denoting $2^{\aleph_1}$ in $V[G_\beta]$.

Let $\pi: \kappa \to \kappa \times \kappa$ be a pairing function as in Section 3.

Now if α is even, say $\alpha = 2\cdot\alpha'$, let $\pi(\alpha') = (\beta,\gamma)$. Let $\dot{Q}_\alpha$ be a term such that $\dot{Q}_\alpha$ denotes the same ordering Q in $V[G_\alpha]$ that $\dot{Q}^\gamma_\beta$ denotes in $V[G_\beta]$, provided Q has the countable chain condition in $V[G_\alpha]$; otherwise $\dot{Q}_\alpha$ denotes the trivial ordering.

If α is odd, say $\alpha = 2\cdot\alpha' + 1$, let $\pi(\alpha') = (\beta,\gamma)$. Let $\dot{Q}_\alpha$ be a term such that if T is the tree denoted by $\dot{T}^\gamma_\beta$ in $V[G_\beta]$ then $\dot{Q}_\alpha$ denotes $P(T)$; if $\dot{T}^\gamma_\beta$ is undetermined (i.e., if $\gamma \geq$ the denotation of $\dot{\lambda}_\beta$ in $V[G_\beta]$) then $\dot{Q}_\alpha$ denotes the trivial ordering.

Since $|P_\alpha| < \kappa$ and κ is inaccessible, it is straightforward to check that $\Vdash_\alpha |\dot{Q}_\alpha| < \kappa$. Clearly $\Vdash_\alpha \dot{Q}_\alpha$ satisfies Axiom A. Hence by Theorem 7.1, P_κ does not collapse ω_1. One checks that MA holds in $V[G_\kappa]$ for orderings of cardinality $< \kappa$ just as in Section 3.

To see that $\Vdash_\kappa$ "Every $\aleph_1$-tree is special", argue first as in Lemma 3.5 that if $T \in V[G_\kappa]$ and T is an $\aleph_1$-tree then $(\exists\beta < \kappa)\ T \in V[G_\beta]$. (This requires the κ-chain condition.) Hence T is the denotation of some $\dot{T}^\gamma_\beta$. But if $\pi(\alpha') = (\beta,\gamma)$, then at stage $\alpha = 2\cdot\alpha' + 1$, T was made special. And, of course, once T was made special it remained special in all further extensions.

Since $\aleph_1$ and κ both remain cardinals in $V[G_\kappa]$, we have $\kappa \geq \aleph_2^{V[G_\kappa]}$. We claim $\kappa = \aleph_2^{[G_\kappa]}$. If not, let $\lambda = \aleph_2^{V[G_\kappa]}$. Since $\lambda < \kappa$, one sees as in Lemma 3.5 that for some $\beta < \kappa$, $V[G_\beta]$ contains onto mappings $f: \omega_1 \to \mu$ for all $\mu < \lambda$. But then $\lambda = \aleph_2^{V[G_\beta]}$. If T is an $\aleph_1$-tree in $V[G_\beta]$, then at some later point T is made special by forcing with $P(T)$. But among other things P(T) collapses $2^{\aleph_1}$ onto $\aleph_1$. Hence λ is collapsed onto $\aleph_1$, contradiction. Thus $\kappa = \aleph_2^{V[G_\kappa]}$.

The proof that $2^{\aleph_0} = \kappa$ in $V[G_\kappa]$ is left to the reader. This completes the proof. □

Remarks. 1. Silver [19] showed that in the model obtained by collapsing all cardinals below an inaccessible onto $\aleph_1$, CH holds and there are no Kurepa trees. However, CH implies that Canadian trees exist; just consider the complete binary tree of height ω_1. Silver also showed that in order to obtain a model with no Kurepa trees it must be consistent that an inaccessible cardinal exists. Hence the inaccessible is needed in Theorem 8.8.

Mitchell [14] showed the consistency (assuming an inaccessible) of the assertion that there are no Canadian trees. Our argument, particularly Theorems 8.5 and 8.6, bears the imprint of Mitchell's approach.

Finally, K. Devlin [4] showed with a long and difficult argument that by forcing over Silver's model with an ordering as in Section 3, a model of MA + $2^{\aleph_0} = \aleph_2$ + "There are no Kurepa trees" results. Unfortunately, there are Canadian trees in Devlin's model since any Canadian tree in Silver's model remains Canadian in Devlin's model.

2. In recent work, Shelah and the author have found several strengthened versions of MA for proper and Axiom A forcing. These stronger axioms imply that every $\aleph_1$-tree is special, as well as other results that do not follow from ordinary MA, but to prove the axioms consistent some fairly large cardinals seem to be needed. These axioms will therefore be discussed elsewhere.

9. Iterated Mathias forcing and the Borel Conjecture.

A set X of real numbers is said to have strong measure zero iff for any sequence $\langle \varepsilon_n : n \in \omega \rangle$ of positive real numbers there is a sequence $\langle I_n : n \in \omega \rangle$ of intervals such that the diameter of I_n is $\leq \varepsilon_n$ and $X \subseteq \cup\{I_n : n \in \omega\}$. The Borel Conjecture asserts that every set of strong measure zero is countable.

In [11], Laver proved the consistency of the Borel Conjecture using iterated Laver forcing with countable support. He remarked that the same could be done with iterated Mathias forcing; that is the goal of this section. For a more complete discussion of the Borel Conjecture, see [11].

The Mathias ordering P consists of all pairs (s,A), where s is a finite subset of ω, A is an infinite subset of ω and $(\forall m \in s)(\forall n \in A)m < n$. Let $(s,A) \leq (t,B)$ iff $s \supseteq t$, $A \subseteq B$ and $s - t \subseteq B$. For a thorough discussion of Mathias forcing, see [12]. First we argue that P satisfies Axiom A.

For $n \geq 1$, let $(s,A) \leq_n (t,B)$ iff $(s,A) \leq (t,B)$, $s = t$ and the first n elements of B belong to A. Of course, $(s,A) \leq_0 (t,B)$ iff $(s,A) \leq (t,B)$.

It is clear that if $\langle (s_n,A_n) : n \in \omega \rangle$ is a sequence such that $(s_{n+1},A_{n+1}) \leq_n (s_n,A_n)$ for all n, then $(s_1,\cap\{A_m : n \in \omega\}) \leq_n (s_n,A_n)$ for all n. It remains only to check (4) (or, equivalently, (4')) in the definition of Axiom A.

For simplicity we use $\Vdash$ instead of $\Vdash_P$ in the following results.

Theorem 9.1. If $(s,A) \Vdash \mathring{a} \in V$, then there is $B \subseteq A$ and countable $x \in V$ such that $(s,B) \Vdash \mathring{a} \in x$. Moreover, B may be chosen so that if $(t,C) \leq (s,B)$, $a \in V$ and $(t,C) \Vdash \mathring{a} = a$, then $(t,B-(\max(t)+1)) \Vdash \mathring{a} = a$.

Proof. We will construct inductively a sequence $b_0 < b_1 < b_2 < \ldots$ of elements of A and a sequence $B_0 \supseteq B_1 \supseteq B_2 \supseteq \ldots$ of subsets of A such that $(\forall b \in B_{n+1}) b_n < b$. Let $B_0 = A$. Given B_n, let $s_1,\ldots,s_k$ enumerate all the subsets of $\{b_i : i < n\}$. Now construct a sequence $B_0^n \supseteq B_1^n \supseteq \ldots \supseteq B_k^n$ as follows. Let $B_0^n = B_n$. Given B_i^n, if there exists $C \subseteq B_i^n$ such that for some $a \in V$, $(s \cup s_i, C) \Vdash \mathring{a} = a$, then let B_{i+1}^n be such a set C; if not let $B_{i+1}^n = B_i^n$. Finally let b_n be the least element of B_k^n and let $B_{n+1} = B_k^n - \{b_n\}$. Let $B = \{b_n : n \in \omega\}$.

Let $x = \{a \in V: (\exists t \subseteq B)(s \cup t, B-(\max(t)+1)) \Vdash \dot{a} = a\}$. Then x is countable, and we claim $(s,B) \Vdash \dot{a} \in x$. Suppose $(t,C) \le (s,B)$ and $(t,C) \Vdash \dot{a} = a$. Let $t - s \subseteq \{b_i : i < n\}$. Using the notation for step n in the induction above, assume $t - s = s_i$. Since $(t,C) \Vdash \dot{a} = a$, we must have chosen B^n_{i+1} so that for some a', $(s \cup s_i, B^n_{i+1}) \Vdash \dot{a} = a'$. But $(t,C) \le (s \cup s_i, B^n_{i+1}) = (t, B^n_{i+1})$, so $a = a'$. But also $(t, B-(\max(t)+1)) \le (t, B^n_{i+1})$ so $a \in x$ and we are done. □

Corollary 9.2. P satisfies (4') in the definition of Axiom A.

Proof. Suppose $(s,A) \Vdash \dot{a} \in V$. Let n be fixed and let t consist of the first n elements of A. Let $t_1, \ldots, t_k$ enumerate the subsets of t. Determine $B_0 \supseteq B_1 \supseteq \ldots \supseteq B_k$ as follows. Let $A = B_0$. Given B_i, apply Theorem 9.1 to find $B_{i+1} \subseteq B_i$ and countable $x_{i+1} \in V$ such that $(s \cup t_i, B_{i+1}) \Vdash \dot{a} \in x_{i+1}$. Let $B = t \cup B_k$. If $x = x_1 \cup x_2 \cup \ldots \cup x_k$, then it is routine to see that $(s,B) \Vdash \dot{a} \in x$. □

Thus the Mathias ordering satisfies Axiom A. There is one additional property, however, which is crucial for the Borel Conjecture.

Theorem 9.3. Let ϕ be a sentence of the language of forcing. For any $(s,A) \in P$ there is $B \subseteq A$ such that either $(s,B) \Vdash \phi$ or $(s,B) \Vdash \neg\phi$.

Proof. First note that there exists $B' \subseteq A$ such that if $(t,C) \le (s,B')$ and $(t,C) \Vdash \phi$ (resp. $\neg\phi$), then $(t, B'-(\max(t)+1)) \Vdash \phi$ (resp. $\neg\phi$). This may be deduced from Theorem 9.1 using the following trick. Let $\dot{a}$ be a term which denotes 0 if ϕ holds in V[G] and which denotes 1 if $\neg\phi$ holds in V[G]. Then $\Vdash \dot{a} \in \{0,1\}$ and $(\forall p \in P) p \Vdash \dot{a} = 0$ iff $p \Vdash \phi$ (and similarly for $\neg\phi$).

For the purposes of this proof, if $t \subseteq B'$, let $t \Vdash \phi$ abbreviate $(s \cup t, B'-(\max(t)+1)) \Vdash \phi$, and similarly for $\neg\phi$.

Now we construct elements $b_0 < b_1 < \ldots$ of B' and subsets $B_0 \supseteq B_1 \supseteq \ldots$ of B' by induction as follows. Let $B_0 = B'$. Given B_n find $B'_{n+1} \subseteq B_n$ so that for all $s' \subseteq \{b_i : i < n\}$ one of the following alternatives holds:

(1) $(\forall b \in B'_{n+1})\ s' \cup \{b\} \Vdash \phi$

(2) $(\forall b \in B'_{n+1})\ s' \cup \{b\} \Vdash \neg \phi$

(3) $(\forall b \in B'_{n+1})$ neither $s' \cup \{b\} \Vdash \phi$ or $s' \cup \{b\} \Vdash \neg \phi$.

Let b_n be the least element of B'_{n+1}, and let $B_{n+1} = B'_{n+1} - \{b_n\}$. Let $B = \{b_n : n \in \omega\}$.

Now suppose $(t,C) \le (s,B)$ and $(t,C) \Vdash \phi$. (The case for $\neg \phi$ is exactly similar.) Let $|t|$ be minimal. If $|t| = |s|$ then $t = s$ and $(s,B) \Vdash \phi$ by the assumption on B'. If $|t| > |s|$ then $\max(t) = b_n$ for some n. Then at stage n we must have had (1) above for $s' = t - (s \cup \{b_n\})$. But then we would have had $(s \cup s', B_{n+1}) \Vdash \phi$ so $(s \cup s', C) \Vdash \phi$ and $|s \cup s'| < |t|$, contradicting minimality of $|t|$. □

<u>Corollary 9.4</u>. If $x \in V$ is finite and $(s,a) \Vdash \mathring{a} \in x$, then for any $n \in \omega$ there is $(t,B) \le_n (s,A)$ and $y \subseteq x$ such that $|y| \le 2^n$ and $(t,B) \Vdash \mathring{a} \in y$. There is also $B \subseteq A$ and $a \in x$ such that $(s,B) \Vdash \mathring{a} = a$.

<u>Proof</u>. For $n = 0$ this is trivial. Repeated use of Theorem 9.3 shows that if $(s,A) \Vdash \mathring{a} \in x$ then there is $B \subseteq A$ and $a \in x$ such that $(s,B) \Vdash \mathring{a} = a$. (Choose $a \in x$ and let ϕ be the assertion that $\mathring{a} = a$. If $B \subseteq A$ and $(s,B) \Vdash \phi$ we are done; if $(s,B) \Vdash \neg \phi$ then $(s,B) \Vdash \mathring{a} \in x - \{a\}$, and we proceed inductively.) Now fix $n \ge 1$. Let u consist of the first n elements of A. Let $s_1, \ldots, s_k$ $(k = 2^n)$ enumerate the subsets of u. Determine $B_0 \supseteq B_1 \supseteq \ldots \supseteq B_k$ as follows. Let $B_0 = A - u$. Given B_i, use the observation above to find $B_{i+1} \subseteq B_i$ and $a_{i+1} \in x$ such that $(s \cup s_{i+1}, B_{i+1}) \Vdash \mathring{a} = a_{i+1}$. Let $B = B_k \cup u$ and let

$y = \{a_1, a_2, \ldots, a_k\}$. Then it is easy to see that $(s,B) \Vdash \dot{a} \in y$, and of course $(s,B) \leq_n (s,A)$. □

From now on, let P_α denote an α-stage iteration with countable support (inverse limits at ordinals of cofinality ω; direct limits elsewhere) such that for all $\beta < \alpha$,

$$\Vdash_\beta \dot{Q}_\beta \text{ is the Mathias ordering.}$$

We need some technical lemmas before we can prove the theorem. The notation of Section 7 will be used.

Lemma 9.5. Suppose $x \in V$ is countable and $p \Vdash_\alpha \dot{a} \in x$. For any finite $F \subset \alpha$ and any n, there is $q \leq_{F,n} p$ and $y \subseteq x$ such that $|y| < 2^{n|F|}$ and $q \Vdash_\alpha \dot{a} \in y$.

Proof. The proof is by induction on α.

First suppose α is limit. Fix β such that $F \subseteq \beta < \alpha$. There must be terms $\dot{b}$ and $\dot{f}$ of the language of forcing with P_β so that

$$p|\beta \Vdash_\beta \text{"}\dot{b} \in x,\ \dot{f} \in P_{\beta\alpha},\ \dot{f} \leq p^\beta \text{ and } \dot{f} \Vdash \dot{a} = \dot{b}.\text{"}$$

(The last forcing symbol refers to forcing with $P_{\beta\alpha}$.) Now by Lemma 7.3(c) and the inductive hypothesis, there exist $q' \leq_{F,n} p|\beta$, $f \in P_{\beta\alpha}$ and $y \subseteq x$ such that $|y| \leq 2^{n|F|}$ and $q' \Vdash_\beta \dot{b} \in y$ and $\dot{f} = f$. But then $q' \cup f \leq_{F,n} p$ and $q' \cup f \Vdash_\alpha \dot{a} \in y$.

Now suppose $\alpha = \beta + 1$. If $\beta \notin F$ then proceed as in the case above. So assume $\beta \in F$. By Corollary 9.4 there exist $\dot{q}$ and $\dot{y}_1, \ldots, \dot{y}_k$, where $k = 2^n$, such that

$$p|\beta \Vdash_\beta \text{"}\dot{y}_1, \ldots, \dot{y}_k \in x,\ \dot{q} \leq_n p(\beta) \text{ and } \dot{q} \Vdash \dot{a} \in \{\dot{y}_1, \ldots, \dot{y}_k\}\text{"}.$$

(Here, of course, the last forcing symbol refers to forcing with $\dot{Q}_\beta$.) By inductive hypothesis there exist $q' \leq_{F\cap\beta,n} p(\beta)$ and $y_1, \ldots, y_k \subseteq x$ such that $|y_i| \leq 2^{n(|F|-1)}$, $i = 1,2,\ldots,k$, and

$$q' \Vdash_\beta \dot{y}_1 \in y_1, \ldots, \dot{y}_k \in y_k.$$

But now if $y = y_1 \cup \ldots \cup y_k$ then $|y| \leq 2^{n|F|}$ and if $q|\beta = q'$ and $q(\beta) = \dot{q}$ then $q \leq_{F,n} p$ and $q \Vdash_\alpha \dot{a} \in y$. □

Lemma 9.6. Suppose $\langle x_n : n \in \omega \rangle \in V$ is a sequence of finite sets. If $p \Vdash_\alpha \forall n f(n) \in x_n$, then there exist $q \leq p$ and a sequence $\langle y_n : n \in \omega \rangle \in V$ such that for all n, $y_n \subseteq x_n$ and $|y_n| \leq 2^{n^2}$, and $q \Vdash_\alpha \forall n \dot{f}(n) \in y_n$.

Proof. Use Lemma 9.5 repeatedly to construct $\langle y_n : n \in \omega \rangle$ and a fusion sequence $\langle (p_n, F_n) : n \in \omega \rangle$ such that $p_0 = p$, $|F_n| = n$, and $p_{n+1} \Vdash_\alpha \dot{f}(n) \in y_n$. If $q \leq_{F_n,n} p_n$ for all n, then $q \Vdash_\alpha \forall n \dot{f}(n) \in y_n$. □

Now we are ready for the main theorem.

Theorem 9.7. Assume CH. Then $\Vdash_{\omega_2}$ The Borel Conjecture.

Proof. By Theorem 7.1 we know that P_{ω_2} preserves cardinals and has the $\aleph_2$-chain condition. We leave as an exercise the verification that $\Vdash_{\omega_2} 2^{\aleph_0} = \aleph_2$.

In order to treat the Borel Conjecture, it will be convenient to reduce it to a proposition about the space ${}^{\omega}2$ with the product measure generated by the measure on 2 giving {0} and {1} both measure $\frac{1}{2}$. It is easy to see that the union of countably many sets of reals of strong measure zero still has strong measure zero, so it suffices to prove the Borel Conjecture for subsets of the unit interval [0,1]. Moreover, the mapping ϕ which associates to each real in [0,1] the function in ${}^{\omega}2$ obtained from its binary representation has the property that ϕ is one-to-one, its range is all but a countable subset of ${}^{\omega}2$, and ϕ is measure-preserving. Let us say that $X \subseteq {}^{\omega}2$ has strong measure zero iff $\phi^{-1}(X)$ has strong measure zero. Hence it suffices to prove the Borel Conjecture for subsets of ${}^{\omega}2$.

Now ${}^{\omega}2$ has a basis consisting of Baire intervals of the form $[s] = \{f \in {}^{\omega}2 : s \subseteq f\}$, where $s \in \cup\{{}^{n}2 : n \in \omega\}$. If $s \in {}^{n}2$

then the measure of $[s]$ is $\frac{1}{2^n}$. With this in mind it should be easy to see that a set $X \subseteq {}^{\omega}2$ has strong measure zero iff $(\forall f \in {}^{\omega}\omega)(\exists g)(\forall n) g(n) \in {}^{f(n)}2$ and $(\forall h \in X)(\exists n) g(n) \subseteq h$. The latter statement is sufficiently set theoretical for our purposes.

Suppose $X \in V[G_{\omega_2}]$ and X is an uncountable subset of ${}^{\omega}2$. Arguing as in Lemma 3.5 and using the $\aleph_2$-chain condition, we see that there must be $\beta < \omega_2$ such that $X \in V[G_{\beta}]$. Defining K_{β} as in Theorem 5.4, we see that a set is K_{β}-thin iff it is countable. Hence by Lemma 7.3(b) the hypothesis of Theorem 5.4 is satisfied, and it follows that in $V[G_{\beta}]$, $P_{\beta\omega_2}$ is (isomorphic to) exactly the same kind of iteration as P_{ω_2}. To be precise, $P_{\beta\omega_2}$ is isomorphic to P_{ω_2} <u>as defined in</u> $V[G_{\beta}]$. Thus without loss of generality we may assume $X \in V$.

Now since P_1 is canonically isomorphic to the Mathias ordering P, the P_1-generic set G_1 determines a P-generic set G. Let $M \in {}^{\omega}\omega$ enumerate $\cup\{s: (\exists A)(s,A) \in G\}$ in increasing order. For each $n \in \omega$, define $M_n \in {}^{\omega}\omega$ by letting $M_n(m) = M(m+n)$. We assert that there exists n such that in $V[G_{\omega_2}]$ there is no g such that $\forall m\ g(m) \in {}^{M_n(m)}2$ and $\forall h \in X\ \exists m\ g(m) \subseteq h$. Now by Lemma 9.6 applied in $V[G_1]$ (and using the fact remarked above that over $V[G_1]$, P_{1,ω_2} is the same kind of iteration as P_{ω_2}), for any such function g there is $y \in V[G_1]$ such that $\forall m\ g(m) \in y(m)$ and $|y(m)| \leq 2^{m^2}$.

Let $Z_n = \{y \in V[G_1]: \forall m\ y(m) \subseteq {}^{M_n(m)}2 \text{ and } |y(m)| \leq 2^{m^2}\}$.

Hence it will suffice to prove in $V[G_1]$ that for some n,

(*) $(\forall y \in Z_n)(\exists h \in X)(\forall m)(\forall s \in y(m))\ s \not\subseteq h$.

This, of course, will involve only forcing with the Mathias ordering P.

Since X is uncountable it is easy to see that for each $m \in \omega$ there is $k(m) \in \omega$ such that

$$|\{s \in {}^{k(0)}2\colon [s] \cap X \text{ is uncountable}| \geq 2 \text{ and}$$

for all m, if $s \in {}^{k(m)}2$ and $[s] \cap X$ is uncountable then $|\{t \in {}^{k(m+1)}2\colon s \subseteq t \text{ and } [t] \cap X \text{ is uncountable}\}| > 2^{(m+1)^2}$. Note that if $y \in Z_n$ and $j \in \omega$ then there are uncountably many $h \in X$ such that $(\forall i \leq j)(\forall s \in y(i))s \nsubseteq h$.

Now let $(s,A) \in P$ be arbitrary. Let $n = |s|$ and let $B \subseteq A$ be such that if $N \in {}^{\omega}\omega$ enumerates B in increasing order, then $N(m) \geq k(m)$ for all m. We have chosen n minimal so that (s,A) decides no value of M_n. We assert that $(s,B) \Vdash (*)$, and this will complete the proof.

Suppose on the contrary that $(s,B) \nVdash (*)$. Then there exist $(t,C) \leq (s,B)$ and $\dot{y}$ such that

$$(t,C) \Vdash \dot{y} \in \dot{Z}_n \text{ and } (\forall h \in X)(\exists m)(\exists s' \in \dot{y}(m))s' \subseteq h.$$

Here $\dot{Z}_n$ is the name for Z_n.

<u>Lemma 9.8</u>. There is $D \subseteq C$ such that for every finite $u \subseteq D$ and every m such that $m + n < |t \cup u|$, there is $z = z(m,u)$ such that $(t \cup u, D-(\max(u)+1)) \Vdash \dot{y}(m) = z$.

<u>Proof</u>. Note that the condition on m and u guarantees that $(t \cup u, C-(\max(u)+1))$ decides $\dot{M}_n(m)$ and hence restricts the set of possibilities for $\dot{y}(m)$ to a finite set. We construct elements $d_0 < d_1 < \ldots$ of C and subsets $D_0 \supseteq D_1 \supseteq \ldots$ of C as follows. Let $D_0 = C$. Given D_j, let $(u_0,m_0),\ldots,(u_k,m_k)$ enumerate all pairs (u,m) such that $m + n < |t \cup u|$ and $u \subseteq \{d_i\colon i < j\}$. We obtain $D_0^j \supseteq D_i^j \supseteq \ldots \supseteq D_k^j$ as follows. Let $D_0^j = D_j$. Given D_i^j, note that for some $r \in \omega$, $(t \cup u_i, D_i^j) \Vdash \dot{M}_n(m_i) = r$ and $\dot{y}(m_i) \subseteq {}^{r}2$. Hence by Corollary 9.4 there is $D_{i+1}^j \subseteq D_i^j$ and z such that $(t \cup u_i, D_{i+1}^j) \Vdash \dot{y}(m_i) = z$. Let d_j be the least element of D_k^j, and let $D_{j+1} = D_k^j - \{d_j\}$. Finally, let $D = \{d_i\colon i \in \omega\}$.

It is easy to check that D works. □

Now if $h \in X$ and u is a finite subset of D, let us call h u-good iff $(\forall u' \subseteq u)(\forall m)$ if $m + n < |t \cup u'|$ then $\forall v \in z(m,u')\ v \not\subseteq h$. Call h avoidable iff $\forall u \subseteq D$ if h is u-good then $\exists u' \subseteq D\ u \subsetneq u'$ and h is u'-good.

If there is $h \in X$ which is avoidable and 0-good, then we may construct a sequence $0 = u_0 \subsetneq u_1 \subsetneq \ldots$ such that h is u_i-good for all i. But if $E = \cup\{u_i : i \in \omega\}$ then $(t,E) \Vdash (\forall m)(\forall v \in \dot{y}(m))v \not\subseteq h$, and this contradicts our original assumption about (t,C).

Now if m is such that $m + n < |t|$ then by Lemma 9.8, (t,D) decides the value of $\dot{y}(m)$. Hence by the remark following the definition of k(m) there must be uncountably many $h \in X$ which are 0-good. Since none of these 0-good h can be avoidable, there must be $u \subseteq D$ such that if $K = \{h \in X$: h is u-good but not $u \cup \{k\}$-good for any $k \in D - u\}$, then K is uncountable. Note that if $h \in K$, $k \in D$ and $u \subseteq k$, then we must have $h|k \in z(u' \cup \{k\}, m(u'))$ for some $u' \subseteq u$, where $m(u')$ is chosen so that $m(u') + n = |t \cup u'|$.

Let $\ell > 2^{2m(u)^2}$, and let $h_1,\ldots,h_\ell$ be distinct members of K. Choose $k \in D$ so large that $u \subseteq k$ and $h_1|k,\ldots,h_\ell|k$ are distinct. Now $h_1|k,\ldots,h_\ell|k \in \cup\{z(u' \cup \{k\}, m(u')) : u' \subseteq u\}$. On the other hand, $|z(u' \cup \{k\}, m(u'))| \le 2^{m(u')^2} \le 2^{m(u)^2}$. Hence $|\cup\{z(u' \cup \{k\}, m(u')) : u' \subseteq u\}| \le 2^{|u|}2^{m(u)^2} < 2^{2m(u)^2} < \ell$; contradiction. This completes the proof. □

Remarks 1. It is clear from the proof of Theorem 9.7 that it is not necessary to iterate Mathias forcing at every stage. It will suffice to have a cofinal set of α such that

$$\Vdash_\alpha \dot{Q}_\alpha \text{ is the Mathias ordering}$$

provided that for every $\alpha < \omega_2$ we have

$$\Vdash_\alpha \text{"}|\dot{Q}_\alpha| = \aleph_1 \text{ and } \dot{Q}_\alpha \text{ satisfies the first sentence of Corollary 9.4."}$$

Examples of orderings satisfying Corollary 9.4 are Laver forcing, Sacks forcing, and Prikry-Silver forcing. (See Section 7 for the definitions.)

2. One naturally asks whether it is possible to obtain the consistency of the Borel Conjecture with $2^{\aleph_0} = \aleph_3$. This problem is still open, and it is instructive to see where the obvious proof (an iteration of length ω_3 with countable support) breaks down. Roitman was the first to observe that in many iterations cardinals are collapsed for reasons having nothing to do with the combinatorial properties of the component orderings; this is a minor variation on her observation.

Suppose $2^{\aleph_0} > \aleph_1$ and P_{ω_1} is an ω_1-stage iteration with countable support such that for all α,

(**) $\Vdash_\alpha \dot{Q}_\alpha$ has at least two incompatible elements.

Then forcing with P_{ω_1} collapses $2^{\aleph_0}$ onto $\aleph_1$. The proof goes as follows. For each α, choose $\dot{p}_\alpha$, $\dot{q}_\alpha$ so that

$$\Vdash_\alpha \dot{p}_\alpha, \dot{q}_\alpha \text{ are incompatible elements of } \dot{Q}_\alpha.$$

Now, in $V[G_{\omega_1}]$ define for each $\alpha < \omega_1$ a set $S_\alpha = \{n \in \omega$: the interpretation of $\dot{p}_{\omega\cdot\alpha+n}$ in $V[G_{\omega\alpha+n}]$ lies in the $\dot{Q}_\alpha$-generic set G obtained from $G_{\omega\alpha+n+1}\}$. An elementary denseness argument using the countable support of P_{ω_1} shows that every subset of ω in V appears as some S_α.

In the case of iterated Mathias forcing, $2^{\aleph_0} = \aleph_2$ in $V[G_{\omega_2}]$ so the argument above shows that $\aleph_2$ is collapsed onto $\aleph_1$ in $V[G_{\omega_2+\omega_1}]$. Hence an ω_3-stage iteration still yields a model of $2^{\aleph_0} = \aleph_2$! This problem occurs with many other iterations as well.

Note, by the way, that our iterations in Sections 3, 4, and

8 do not satisfy (**), since there may be $p \in P_\alpha$ such that

$$p \Vdash_\alpha \dot{Q}_\alpha \text{ is the trivial ordering.}$$

Nevertheless, (**) can be weakened to include these iterations as well. This yields another proof that $2^{\aleph_0} = \aleph_2$ in the model of Section 8.

References

1. J. Baumgartner and R. Laver, Iterated perfect-set forcing, Annals Math. Logic 17 (1979), 271-288.

2. J. Baumgartner, J. Malitz and W. Reinhardt, Embedding trees in the rationals, Proc. Nat. Acad. Sci. U.S.A. 67 (1970), 1748-1753.

3. P.J. Cohen, Set Theory and the Continuum Hypothesis, Benjamin, New York, 1966.

4. K.J. Devlin, $\aleph_1$-Trees, Annals Math. Logic 13 (1978), 267-330.

5. K.J. Devlin and H. Johnsbråten, The Souslin problem, Lecture Notes in Mathematics 405, Springer-Verlag, Berlin, 1974.

6. W.B. Easton, Powers of regular cardinals, Annals Math. Logic 1 (1970), 139-178.

7. F. Galvin, Chain conditions and products, to appear.

8. C.D. Herink, Ph.D. Dissertation, University of Wisconsin, 1978.

9. T. Jech, Set Theory, Academic Press, New York, 1978.

10. K. Kunen, Some applications of iterated ultrapowers in set theory, Annals Math. Logic 1 (1970), 179-227.

11. R. Laver, On the consistency of Borel's conjecture, Acta Math. 137 (1976), 151-169.

12. A.R.D. Mathias, Happy families, Annals Math. Logic 12 (1977), 59-111.

13. T.K. Menas, Consistency results concerning supercompactness, Trans. Amer. Math. Soc. 223 (1976), 61-91.

14. W. Mitchell, Aronszajn trees and the independence of the transfer property, Annals Math. Logic 5 (1972), 21-46.

15. S. Shelah, A weak generalization of MA to higher cardinals, Israel J. Math. 30 (1978), 297-306.

16. S. Shelah, notes on the P-point problem.

17. S. Shelah, handwritten notes on proper forcing, 1978.

18. J.R. Shoenfield, Unramified forcing, in Axiomatic Set Theory, D. Scott, Editor, Proc. Symp. Pure Math. 13 (1), Amer. Math. Soc. Providence, R.I., 1971, 357-381.

19. J. Silver, The independence of Kurepa's conjecture and two cardinal conjectures in model theory, in Axiomatic Set Theory, D. Scott, editor, Proc. Symp. Pure Math. 13 (1), Amer. Math. Soc., Providence, R.I., 1971, 383-390.

20. J. Silver, unpublished notes on reverse Easton forcing, 1971.

21. R.M. Solovay and S. Tennenbaum, Iterated Cohen extensions and Souslin's problem, Ann. Math. 94 (1971), 201-245.

Dartmouth College,
Hanover, New Hampshire, U.S.A. 03755.

THE YORKSHIREMAN'S GUIDE TO PROPER FORCING

Keith J. Devlin

University of Lancaster, U.K.

§0. INTRODUCTION

Properness is a property of forcing posets which generalises the countable chain condition, is preserved under countable support iterations, and which, when applied to iterations of posets of cardinality $\aleph_1$ (the posets one usually wants to iterate) guarantees the preservation of all cardinals. The notion was formulated by Saharon Shelah, and arose from his study of Ronald Jensen's technique of iterated Souslin forcing (see [DeJo]) and from his attempts (eventually successful) to show that the GCH does not resolve the Whitehead Problem (see [Ek], together with §4 of these notes). In order to motivate the notion, let us review some basic forcing theory, which we will in any case need later.

If $\mathbb{P}$ is a poset (we always assume that $\mathbb{P}$ has a maximum element, $\mathbb{1}$, and that for each $p \in \mathbb{P}$ there are $q,r \in \mathbb{P}$ such that

$q \leq p$, $r \leq p$, and q,r are incomparable in $\mathbb{P}$), there is a uniquely defined complete boolean algebra BA($\mathbb{P}$) which extends $\mathbb{P}$ as a partially ordered set and has $\mathbb{P}$ as a dense subset (in the usual, poset sense). We denote by $V^{(\mathbb{P})}$ the boolean universe $V^{(BA(\mathbb{P}))}$ (see [Je] for basic forcing theory). To simplify arguments, it is common to speak of a "V-generic subset, G, of $\mathbb{P}$" and thence of "the extension V[G] of V". Though technically wrong, this procedure is easily eliminated by working entirely with countable models of set theory, but it is more convenient to avoid this restriction, so we shall make frequent use of the above mentioned abuse. In this connection, each element, x, of V[G] is the "collapse" of some (in fact many) boolean set of $V^{(\mathbb{P})}$. Any member of $V^{(\mathbb{P})}$ which collapses to x in V[G] is called a <u>$\mathbb{P}$-name</u> of x. (Shoenfield, in [Sh], makes use of a canonical collection of $\mathbb{P}$-names in order to develop his theory of unramified forcing.) We often use $\overset{\circ}{x}$ to denote a $\mathbb{P}$-name of $x \in V[G]$. Elements of V, which are <u>a fortiori</u> elements of V[G], have canonical names: the canonical name of $x \in V \subset V[G]$ is denoted by $\check{x}$.

The boolean truth-value of a sentence ϕ of set theory when interpreted in $V^{(\mathbb{P})}$ is denoted by $||\phi||^{\mathbb{P}}$. The forcing relation between members of $\mathbb{P}$ and sentences ϕ is defined by $p \Vdash_{\mathbb{P}} \phi$ iff $p \leq ||\phi||^{\mathbb{P}}$ (in BA($\mathbb{P}$)).

Since cardinalities are not in general preserved when we pass from one universe of set theory to another, we are interested in properties of posets which guarantee the preservation of cardinals.

The most common are as follows.

If $\mathbb{P}$ satisfies the <u>κ chain condition</u> (i.e. $\mathbb{P}$ has no pairwise incompatible subset of cardinality κ), where κ is a regular uncountable cardinal, then $\|\check{\lambda}$ is a cardinal $\|^{\mathbb{P}} = \mathbb{1}$ for all cardinals $\lambda \geq \kappa$. In particular, if $\mathbb{P}$ satisfies the <u>countable chain condition</u> (=c.c.c. = $\aleph_1$ chain condition), then $\|\check{\lambda}$ is a cardinal $\|^{\mathbb{P}} = \mathbb{1}$ for all cardinals λ.

If $\mathbb{P}$ is <u>κ-closed</u> (i.e. every decreasing sequence of elements of $\mathbb{P}$ of length less than κ has an infimum in $\mathbb{P}$), where κ is a regular uncountable cardinal, then for all $\alpha < \kappa$, $\|\mathcal{P}(\check{\alpha}) = \widetilde{\mathcal{P}(\alpha)}\|^{\mathbb{P}} = \mathbb{1}$ (i.e. V and $V^{(\mathbb{P})}$ have the same subsets of α for all $\alpha < \kappa$), so in particular $\|\check{\lambda}$ is a cardinal $\|^{\mathbb{P}} = \mathbb{1}$ for all cardinals $\lambda \leq \kappa$.

If $\mathbb{P}$ is <u>κ-dense</u> (i.e. the intersection of any family of fewer than κ dense initial sections of $\mathbb{P}$ is dense in $\mathbb{P}$), where κ is a regular uncountable cardinal, then for all $\alpha < \kappa$, $\|\mathcal{P}(\check{\alpha}) = \widetilde{\mathcal{P}(\alpha)}\|^{\mathbb{P}} = \mathbb{1}$, so again $\|\check{\lambda}$ is a cardinal $\|^{\mathbb{P}} = \mathbb{1}$ for all cardinals $\lambda \leq \kappa$.

Clearly, if $\mathbb{P}$ is κ-closed then it is automatically κ-dense. The converse is false (very false!).

Suppose $\mathbb{P}$ is a poset, and that in $V^{(\mathbb{P})}$ there is a boolean object $\mathbb{Q}$ such that $\|\mathbb{Q}$ is a poset $\|^{\mathbb{P}} = \mathbb{1}$. We define a poset $\mathbb{P} \otimes \mathbb{Q}$ as follows. As domain we take the set of all pairs (p,q) such that $p \in \mathbb{P}$ and $p \Vdash_{\mathbb{P}}$ "$q \in Q$". (To ensure that this collection is a <u>set</u>, we should identify pairs (p,q), (p,q') such that $p \Vdash_{\mathbb{P}}$ "$q = q'$". In practice we usually ignore this point.) The partial ordering is defined by

$$(p',q') \leq (p,q) \text{ iff } p' \leq_{\mathbb{P}} p \;\&\; p' \Vdash_{\mathbb{P}} "q' \leq_{\mathbb{Q}} q".$$

There is then a canonical structural isomorphism $V^{\mathbb{P} \otimes \mathbb{Q}} \simeq [V^{(\mathbb{P})}]^{(\mathbb{Q})}$. Moreover, if G is V-generic on $\mathbb{P}$ and H is V[G]-generic on $\mathbb{Q}$, then $G \times H$ is V-generic on $\mathbb{P} \otimes \mathbb{Q}$, and every V-generic subset of $\mathbb{P} \otimes \mathbb{Q}$ has the form $G \times H$ for some such G,H. See, for example, [SoTe] for details.

The above "product lemma" provides the key for performing (transfinite) iterations of boolean extensions.

We say that a sequence $\langle \mathbb{P}_\nu \mid \nu \leq \lambda \rangle$ of posets is an <u>iteration sequence</u> if there is a sequence $\langle \mathbb{Q}_\nu \mid \nu < \lambda \rangle$ such that for each $\nu < \lambda$, $\mathbb{Q}_\nu$ is an element of $V^{(\mathbb{P}_\nu)}$ such that $\| \mathbb{Q}_\nu \text{ is a poset} \|^{\mathbb{P}_\nu} = \mathbb{1}$ and $\mathbb{P}_{\nu+1} = \mathbb{P}_\nu \otimes \mathbb{Q}_\nu$. This definition does not tell us anything about $\mathbb{P}_\nu$ in case ν is a limit ordinal, of course, and there are various different requirements we can make to cover this case.

We say that $\langle \mathbb{P}_\nu \mid \nu \leq \lambda \rangle$ is a <u>full iteration sequence</u> if, for each limit ordinal $\nu \leq \lambda$, $\mathbb{P}_\nu$ consists of all sequences $q = \langle q(\zeta) \mid \zeta < \nu \rangle$ such that $q(0) \in \mathbb{P}_0$ and for all $\zeta < \nu$, $q \restriction 1+\zeta \Vdash_{\mathbb{P}_\zeta} "q(1+\zeta) \in \mathbb{Q}_\zeta"$, and the ordering of $\mathbb{P}_\nu$ is

$$q' \leq q \text{ iff } q'(0) \leq q(0) \;\&\; (\forall \zeta < \nu)[q' \restriction 1+\zeta \Vdash_{\mathbb{P}_\zeta} "q'(1+\zeta) \leq q(1+\zeta)"].$$

The <u>support</u> of an element q of $\mathbb{P}_\nu$ is the set of all those $\zeta < \nu$ such that $q(\zeta) \neq \mathbb{1}$. (We use $\mathbb{1}$ as a general symbol for the maximum element of any poset. In case $\zeta > 0$ here, this means that $q(\zeta)$ is not the term : "the maximum element of $\mathbb{Q}_\zeta$".)

The iteration sequence $\langle \mathbb{P}_\nu \mid \nu \leq \lambda \rangle$ is said to have finite support, if, for each limit ordinal $\nu \leq \lambda$, $\mathbb{P}_\nu$ consists of all sequences $q = \langle q(\zeta) \mid \zeta < \nu \rangle$ as above which have finite support, the ordering being defined as above.

Similarly a countable support iteration, which is the type we shall be dealing with in these notes.

The point about an iteration sequence $\langle \mathbb{P}_\nu \mid \nu \leq \lambda \rangle$ is that for any $\nu < \tau \leq \lambda$ there is a canonical element $\mathbb{Q}_\nu^\tau \in V^{(\mathbb{P}_\nu)}$ such that $\mathbb{P}_\tau \approx \mathbb{P}_\nu \otimes \mathbb{Q}_\nu^\tau$. Hence an iteration sequence corresponds to a sequence of successive boolean extensions (of the same length).

Now, if $\langle \mathbb{P}_\nu \mid \nu \leq \lambda \rangle$ is an iteration sequence with finite supports, and if $\mathbb{P}_0$ satisfies the c.c.c. and for each $\nu < \lambda$, $\| \mathbb{Q}_\nu$ satisfies the c.c.c. $\|^{\mathbb{P}_\nu} = 1$, then every poset $\mathbb{P}_\nu$, $\nu \leq \lambda$, satisfies the c.c.c. This is proved in [SoTe]. Thus finite support iterations of c.c.c. posets preserve cardinals. This enabled Solovay and Tennenbaum to construct a model of Martin's Axiom in [SoTe]. A natural question is whether or not there is a more general property than c.c.c. which is preserved by iterations. More specifically, suppose we want to perform an iteration in which not only are cardinals preserved, but CH is preserved as well. Finite support iterations are of no use here, since new real numbers appear at each limit stage. But countable support iterations destroy the c.c.c. Well, if the posets being iterated are both $\aleph_1$-closed and of cardinality $\aleph_1$, then each poset in a countable support iteration is also $\aleph_1$-closed and satisfies the $\aleph_2$-c.c., so such an iteration could preserve both cardinals and the CH. But unfortunately there is not much one can do with $\aleph_1$-closed posets of cardinality $\aleph_1$, as is easily observed! Much more useful would be a countable support iteration of posets of cardinality $\aleph_1$ which are $\aleph_1$-dense. Unfortunately, $\aleph_1$-density is in general not preserved by countable support iterations. So what we require is some property of posets which is more general than $\aleph_1$-closure, which is preserved under countable support iterations, and which implies at the very least that (for posets of size $\aleph_1$, say) cardinals are preserved. These notes discuss just such a property. It turns out to include c.c.c. posets as well: an extra, perhaps unexpected, bonus.

The organisation of this paper is as follows. In §1 we give some basic preliminary results concerning a natural generalisation of the concept of a closed unbounded set of ordinals. §2 introduces the notion of properness, gives some examples of proper posets, and develops alternative characterisations of the notion for later use. In §3 we show that countable support iterations of proper posets are proper (i.e. properness is preserved under countable support iteration.) In §4 we give an illustration of the use of proper forcing by establishing a combinatorial principle which leads to the solution of the Whitehead Problem in Group Theory. In §5 we show how proper forcing can lead to a powerful generalisation of Martin's

Axiom, known as the Proper Forcing Axiom (PFA). Though extremely interesting in its own right, and having many startling consequences (see [Ba]), it should perhaps be stressed that this does not really involve the full potential of proper forcing, which really comes into its own when used (often in more special forms) to perform delicate iterations which do not destroy CH. But that is another story.

Prerequisites? If you have got this far (and followed our outlines) then you must have all of the preliminary knowledge required. If you haven't got this far, then,......!

Finally, if you were present at the Cambridge Meeting of which this is the Proceedings, you may recall that I was not even present, let alone did I give any lectures on Proper Forcing. Indeed, though Shelah was present and did speak about such matters, the whole concept was still in its infancy, and certainly not in the polished (I hope) form presented here, which version I only wrote out in the Autumn of 1980. But the Cambridge Meeting did represent the first "world wide exposure" of proper forcing, and since in the interim no completely worked out account has been made available, the Editor of these Proceedings felt that it was appropriate to include the present paper.

The notes were based first of all upon Shelah's original, rather sketchy notes written in Berkeley in 1978, and then improved considerably following some lectures given by Jim Baumgartner in Toronto in 1980. Stevo Todorcevic pointed out a significant error in what I thought was the final version, and suggested various other improvements. The idea of publishing the notes at all (which I originally wrote in order to sort out the ideas for myself with a view to trying to extend them - and no, I am not saying in what direction!) came independently from Jim Baumgartner, Rudi Göbel, and Stevo Todorcevic.

§1. PRELIMINARIES

For any set A and cardinal κ, we set

$$[A]^{\kappa} = \{X \subseteq A \mid |X| = \kappa\} ,$$

$$[A]^{<\kappa} = \{X \subseteq A \mid |X| < \kappa\} .$$

Let A be an infinite set, and let $C \subseteq [A]^{\aleph_0}$. We say C is unbounded in $[A]^{\aleph_0}$ if for every $X \in [A]^{\aleph_0}$ there is a $Y \in C$ such that $X \subseteq Y$. We say C is closed in $[A]^{\aleph_0}$ if, whenever $X_n \in C$ and $X_n \subseteq X_{n+1}$ for all $n < \omega$, then $\bigcup_{n<\omega} X_n \in C$. As usual, "club" abbreviates "closed and unbounded".

Lemma 1.1

If C_1, C_2 are club subsets of $[A]^{\aleph_0}$, then $C_1 \cap C_2$ is also club in $[A]^{\aleph_0}$.

Proof: Obvious. □

Lemma 1.2

If C_a is a club subset of $[A]^{\aleph_0}$ for each $a \in A$, then the set

$$C = \{X \in [A]^{\aleph_0} \mid (\forall a \in X)(X \in C_a)\}$$

is club in $[A]^{\aleph_0}$.

Proof: Suppose that $X_n \in C$ and $X_n \subseteq X_{n+1}$ for all $n < \omega$. Let $X = \bigcup_{n<\omega} X_n$. We show that $X \in C$. Let $a \in X$. For some $k < \omega$,

$a \in X_k$. For all $n \geq k$, $a \in X_n$: so $X_n \in C_a$. Thus $X = \bigcup_{k \leq n < \omega} X_n \in C_a$. Hence $a \in X \to X \in C_a$, giving $X \in C$. Thus C is closed.

To prove unboundedness now, let $X_0 \in [A]^{\aleph_0}$ be given. Let $\{a_n \mid n<\omega\}$ enumerate X_0. Choose $X_0^0 \in C_{a_0}$ to extend X_0. Then, inductively, choose $X_0^{n+1} \in C_{a_0} \cap C_{a_1} \cap \ldots \cap C_{a_{n+1}}$ to extend X_0^n. (This uses lemma 1.1, of course.) Let $X_1 = \bigcup_{n<\omega} X_0^n$, then $a \in X_0$ implies $X_1 \in C_a$. Repeating this process inductively, pick $X_{n+1} \supseteq X_n$ so that $a \in X_n$ implies $X_{n+1} \in C_a$. Let $X = \bigcup_{n<\omega} X_n$. Clearly, $X \supseteq X_0$ and $X \in C$. □

A set $S \subseteq [A]^{\aleph_0}$ is said to be <u>stationary</u> if it meets every club set $C \subseteq [A]^{\aleph_0}$.

<u>Lemma 1.3</u>

A set $S \subseteq [\omega_1]^{\aleph_0}$ is stationary iff $S \cap \omega_1$ is stationary in ω_1 in the usual sense.

Proof: ($\to$) Let $C \subseteq \omega_1$ be club. We show that $S \cap C \neq \phi$. We may assume that $C \cap \omega = \phi$. Thus if $\alpha \in C$, then $\alpha = \{\beta \mid \beta < \alpha\} \in [\omega_1]^{\aleph_0}$. Clearly, C is club in $[\omega_1]^{\aleph_0}$. Hence $C \cap S \neq \phi$, as required.

($\leftarrow$) Let $C \subseteq [\omega_1]^{\aleph_0}$ be club. Then clearly, $C \cap \omega_1$ is club in ω_1. Hence $(C \cap \omega_1) \cap S \neq \phi$, and again we are done. □

<u>Lemma 1.4.</u>

Let $S \subseteq [A]^{\aleph_0}$ be stationary, and let $f: S \to A$ be such that $f(X) \in X$ for all $X \in S$. Then there is a stationary set $\bar{S} \subseteq S$ such that $f \upharpoonright \bar{S}$ is constant.

Proof: For each $a \in A$, set

$$S_a = \{ X \in S \mid f(X) = a \}.$$

We must show that S_a is stationary for some $a \in A$. Suppose not. Then for each $a \in A$ we can pick a club set $C_a \subseteq [A]^{\aleph_0}$ such that $C_a \cap S_a = \phi$. Let

$$C = \{X \in [A]^{\aleph_0} \mid (\forall a \in X)(X \in C_a)\}.$$

By lemma 1.2, C is club. Hence $C \cap S \neq \phi$. Let $X \in C \cap S$. Since $X \in S$, $f(X)$ is defined, say $f(X) = a$. By the assumptions on f, we have $a \in X$. So as $X \in C$, we have $X \in C_a$. Thus $X \notin S_a$, i.e. $f(X) \neq a$, which is absurd. □

<u>Lemma 1.5</u>

Let A be uncountable, and let $C \subseteq [A]^{\aleph_0}$. The following are equivalent:

(i) C contains a club;

(ii) there is a function $f : [A]^{<\aleph_0} \to [A]^{\aleph_0}$ such that for any set $X \in [A]^{\aleph_0}$, if $f''[X]^{<\aleph_0} \subseteq [X]^{\aleph_0}$, then $X \in C$;

(iii) there is a function $f : [A]^{<\aleph_0} \to A$ such that for any set $X \in [A]^{\aleph_0}$, if $f''[X]^{<\aleph_0} \subseteq X$, then $X \in C$. (We shall say that f <u>concentrates on</u> C in this case.)

Proof: Since the nature of the set A clearly plays no role in this lemma, it suffices to prove it for the case where A is an uncountable cardinal λ.

(i) → (ii). By cutting down C if necessary, we may assume that C is itself club. By induction on n, define $f \restriction [\lambda]^n$ so that for any $X \in [\lambda]^{n+1}$,

$$X \cup [\bigcup_{Y \subset X} f(Y)] \subseteq f(X) \in C.$$

Since C is unbounded, there is no difficulty here. So f is defined on $[\lambda]^{<\aleph_0}$. Suppose that $X \in [\lambda]^{\aleph_0}$ is such that $f''[X]^{<\aleph_0} \subseteq [X]^{\aleph_0}$. Let $\{\alpha_n \mid n<\omega\}$ enumerate X. For each $n<\omega$, we have

$$\{\alpha_i | i<n\} \subseteq f(\{\alpha_i | i<n\}) \subseteq f(\{\alpha_i | i < n+1\}) \subseteq X.$$

But $\text{ran}(f) \subseteq C$ and C is closed in $[\lambda]^{\aleph_0}$. Hence

$$X = \{\alpha_i | i<\omega\} = \bigcup_{n<\omega} f(\{\alpha_i | i<n\}) \in C.$$

(ii)→(iii). Let f be as in (ii). For each $X \in [\lambda]^{<\aleph_0}$, let $\langle f_n(X) \mid n<\omega \rangle$ enumerate f(X) (repetitions allowed). Define $h : [\lambda]^{<\aleph_0} \to \lambda$ by setting

$$h(\{\alpha\}) = \alpha + 1 \quad (\alpha < \lambda)$$

and, for $n>0$ and $\alpha_0 <\ldots< \alpha_n < \lambda$,

$$h(\{\alpha_0,\ldots,\alpha_n\}) = f_i(\{\alpha_0,\ldots,\alpha_j\}),$$

where $n = 2^i.(2j+1)-1$. Suppose that $X \in [\lambda]^{\aleph_0}$ is such that $h''[X]^{<\aleph_0} \subseteq X$. We show that $X \in C$. It suffices to show that $f''[X]^{<\aleph_0} \subseteq [X]^{\aleph_0}$. Let $\{\alpha_0,\ldots,\alpha_m\} \subseteq X$, $\alpha_0<\ldots<\alpha_m$. To show that $f(\{\alpha_0,\ldots,\alpha_m\}) \in [X]^{\aleph_0}$ it suffices to show that $f_k(\{\alpha_0,\ldots,\alpha_m\}) \in X$ for all $k < \omega$. Let $k < \omega$ be given. Set $n = 2^k.(2m+1)-1$. Since $h''[X]^1 \subseteq X$, X has no largest member. So we can find $\alpha_{m+1},\ldots,\alpha_n \in X$ so that $\alpha_m < \alpha_{m+1}<\ldots<\alpha_n$. Then

$$f_k(\{\alpha_0,\ldots,\alpha_m\}) = h(\{\alpha_0,\ldots,\alpha_n\}) \in X,$$

as required.

(iii)→(i). Let f be as in (iii), and set

$$\tilde{C} = \{X \in [\lambda]^{\aleph_0} \mid f''[X]^{<\aleph_0} \subseteq X\}.$$

Then $\tilde{C} \subseteq C$. But $\tilde{C}$ is clearly club in $[\lambda]^{\aleph_0}$. □

Lemma 1.6

Let $\mathfrak{A} = \langle A, \in, \ldots \rangle$ be an uncountable, first-order structure. Then there is a function $f : [A]^{<\aleph_0} \to A$ such that whenever $N \in [A]^{\aleph_0}$ is such that $f''[N]^{<\aleph_0} \subseteq N$, then N is the domain of an elementary substructure of $\mathfrak{A}$.

Proof: It is standard that the set of all $N \in [A]^{\aleph_0}$ which are domains of elementary substructures of $\mathfrak{A}$ is club in $[A]^{\aleph_0}$. Now apply lemma 1.5. □

A function f related to a structure $\mathfrak{A}$ as above will be said to be <u>$\mathfrak{A}$-skolem.</u>

Lemma 1.7

Let λ be an uncountable ordinal, and let $\mathfrak{A} = \langle A, \in, \ldots . \rangle$ be a first-order structure such that $\lambda \subseteq A$. Then the set

$$\{N \cap \lambda \mid N \in [A]^{\aleph_0} \ \& \ \langle N, \in, \ldots \rangle \prec \mathfrak{A}\}$$

is club in $[\lambda]^{\aleph_0}$.

Conversely, if $C \subseteq [\lambda]^{\aleph_0}$ is club, there is a structure $\mathfrak{A} = \langle \lambda, \in, (f_n)_{n<\omega} \rangle$ such that for all $N \in [\lambda]^{\aleph_0}$, if $\langle N, \in, \ldots \rangle \prec \mathfrak{A}$ then $N \in C$.

Proof: The first part is standard. For the second part, let $f : [\lambda]^{<\aleph_0} \to \lambda$ concentrate on C (as in lemma 1.5 (iii)) and let $f_n = f \upharpoonright [\lambda]^n$. □

Lemma 1.8

Let λ, μ be uncountable cardinals, $\lambda < \mu$.

(i) If $C \subseteq [\mu]^{\aleph_0}$ is club, then $C \upharpoonright \lambda = \{X \cap \lambda \mid X \in C\}$ contains a club in $[\lambda]^{\aleph_0}$.

(ii) If $C \subseteq [\lambda]^{\aleph_0}$ is club, then $C[\mu] = \{X \in [\mu]^{\aleph_0} \mid X \cap \lambda \in C\}$ is club in $[\mu]^{\aleph_0}$.

Proof : (i) By lemma 1.5, let $f : [\mu]^{<\aleph_0} \to \mu$ be such that $f''[X]^{<\aleph_0} \subseteq X$ implies $X \in C$ for each $X \in [\mu]^{\aleph_0}$, i.e. let f concentrate on C.

Given $\sigma \in [\mu]^{\leq \aleph_0}$, define, inductively

$$\sigma^0 = \sigma, \; \sigma^{n+1} = \sigma^n \cup f''[\sigma^n]^{<\aleph_0},$$

and set $\hat{\sigma} = \bigcup_{n<\omega} \sigma^n$. Define $g : [\lambda]^{<\aleph_0} \to [\lambda]^{\leq \aleph_0}$ by

$$g(\sigma) = \hat{\sigma} \cap \lambda.$$

The set

$$D = \{X \in [\lambda]^{\aleph_0} \mid g''[X]^{<\aleph_0} \subseteq [X]^{\leq \aleph_0}\}$$

is clearly club in $[\lambda]^{\aleph_0}$.

But if $X \in D$, then $\hat{X} \cap \lambda = X$. Moreover, $f''[\hat{X}]^{<\aleph_0} \subseteq \hat{X}$, so $\hat{X} \in C$.

Thus

$$X = \hat{X} \cap \lambda \in C \restriction \lambda.$$

(ii) Trivial. □

Lemma 1.9

Let λ, μ be uncountable cardinals, $\lambda < \mu$. If $S \subseteq [\lambda]^{\aleph_0}$ is stationary, then

$$S[\mu] = \{X \in [\mu]^{\aleph_0} \mid X \cap \lambda \in S\}$$

is stationary in $[\mu]^{\aleph_0}$.

Proof : By lemma 1.8(i). □

§2. PROPER POSETS

A poset $\mathbb{P}$ is said to be proper if, whenever λ is an uncountable cardinal and S is a stationary subset of $[\lambda]^{\aleph_0}$, then

$$\Vdash_{\mathbb{P}} \text{"}\check{S} \text{ is stationary"}.$$

(Note that $\mathbb{P}$ may collapse λ, or introduce new countable subsets of λ.)

Lemma 2.1

All c.c.c. posets are proper.

Proof: Let $\mathbb{P}$ be a c.c.c. poset, λ an uncountable cardinal, S a stationary subset of $[\lambda]^{\aleph_0}$. Given $\mathring{C} \in V^{(\mathbb{P})}$ such that

$$\Vdash_{\mathbb{P}} \text{"}\mathring{C} \text{ is a club subset of } [\check{\lambda}]^{\aleph_0}\text{"},$$

we must show that

$$\Vdash_{\mathbb{P}} \text{"}\mathring{C} \cap \check{S} \neq \phi\text{"}.$$

Pick $\mathring{f} \in V^{(\mathbb{P})}$ so that

$$\Vdash_{\mathbb{P}} \text{"}\mathring{f} : [\check{\lambda}]^{<\aleph_0} \to \check{\lambda} \text{ concentrates on } \mathring{C}\text{"}.$$

Define $h : [\lambda]^{<\aleph_0} \to [\lambda]^{\aleph_0}$ by setting, for $X \in [\lambda]^{<\aleph_0}$,

$$h(X) = \omega \cup \{\alpha \in \lambda \mid \|\mathring{f}(\check{X}) = \check{\alpha}\|^{\mathbb{P}} > \mathbb{O}\}.$$

(By the c.c.c. for $\mathbb{P}$, h(X) is countable.) Clearly, for any $X \in [\lambda]^{<\aleph_0}$,

$$(*) \qquad \Vdash_{\mathbb{P}} \text{"}\mathring{f}(\check{X}) \in \widecheck{h(X)}\text{"}.$$

Now, the set

$$\{X \in [\lambda]^{\aleph_0} \mid h''[X]^{<\aleph_0} \subseteq [X]^{\aleph_0}\}$$

is club in $[\lambda]^{\aleph_0}$, so we can find an $X \in S$ such that $h''[X]^{<\aleph_0} \subseteq [X]^{\aleph_0}$. By (*) we have

$$\Vdash_{\mathbb{P}} \text{" } \mathring{f}\text{ '' } [\check{X}]^{<\aleph_0} \subseteq \check{X} \text{ "}.$$

Thus

$$\Vdash_{\mathbb{P}} \text{"}\check{X} \in \mathring{C} \cap \check{S}\text{"},$$

as required. □

<u>Lemma 2.2</u>

All countably closed posets are proper.

Proof: Let $\mathbb{P}$ be a countably closed poset, λ an uncountable cardinal, S a stationary subset of $[\lambda]^{\aleph_0}$. Let $\mathring{C} \in V^{(\mathbb{P})}$ be such:

$$\Vdash_{\mathbb{P}} \text{"}\mathring{C} \text{ is a club subset of } [\check{\lambda}]^{\aleph_0}\text{"}.$$

We show that

$$\Vdash_{\mathbb{P}} \text{"}\mathring{C} \cap \check{S} \neq \phi\text{"}.$$

Let $\mathring{f} \in V^{(\mathbb{P})}$ be such that

$$\Vdash_{\mathbb{P}} \text{"}\mathring{f} : [\check{\lambda}]^{<\aleph_0} \to \check{\lambda} \text{ concentrates on } \mathring{C}\text{"}.$$

For each $n < \omega$, let R_n be the relation on $\mathbb{P} \times [\lambda]^{n+1}$ defined by

$$R_n(p,\alpha_0,\alpha_1,\ldots,\alpha_n) \text{ iff } p \Vdash_{\mathbb{P}} \text{"}\mathring{f}(\{\check{\alpha}_1,\ldots,\check{\alpha}_n\}) = \check{\alpha}_0\text{"}.$$

Choose $\kappa \geq \lambda$ regular with $\mathbb{P} \in H_\kappa$, and consider the structure

$$\mathfrak{A} = \langle H_\kappa, \in, \mathbb{P}, (R_n)_{n<\omega} \rangle$$

Given $p \in \mathbb{P}$ we find a $q \leq p$ and a set $X \in [\lambda]^{\aleph_0}$ such that $X \in S$ and

$$q \Vdash_{\mathbb{P}} \text{"}\mathring{f}\text{"}[\check{X}]^{<\aleph_0} \subseteq \check{X}\text{"},$$

which at once implies that

$$\Vdash_{\mathbb{P}} \text{"}(\exists X)(X \in \mathring{C} \cap \check{S})\text{"}.$$

By lemma 1.7 we can find a countable submodel

$$\mathfrak{B} = \langle N, \in, \ldots \rangle \prec \mathfrak{A}$$

such that $p \in N$ and $N \cap \lambda \in S$. Let $X = N \cap \lambda$, and let $\langle \sigma_n | n<\omega \rangle$ enumerate $[X]^{<\aleph_0}$. Since $\mathfrak{B} \prec \mathfrak{A}$, we can pick conditions

$p_n \in \mathbb{P} \cap N$ such that

$$p \geq p_0 \geq p_1 \geq \cdots$$

and for some $\alpha_n \in X$,

$$p_n \Vdash_{\mathbb{P}} \text{"}\mathring{f}(\check{\sigma}_n) = \check{\alpha}_n\text{"}$$

(i.e. $R_m(p_n, \alpha_n, \sigma_n)$, where $m = |\sigma_n|$). Since $\mathbb{P}$ is countably closed we can find a $q \in \mathbb{P}$, $q \leq p_n$ for all $n < \omega$. Then, as required,

$$q \Vdash_{\mathbb{P}} \text{"}\mathring{f}\text{"}[\check{X}]^{<\aleph_0} \subseteq \check{X}\text{"}. \square$$

We shall give a combinatorial characterisation of properness. First some definitions.

Let $\mathbb{P}$ be a poset. A set $D \subseteq \mathbb{P}$ is said to be <u>predense</u> iff, for each $p \in \mathbb{P}$ there is a $q \in D$ such that p and q are compatible. If $p_0 \in \mathbb{P}$, we say D is <u>predense below</u> p_0 iff every extension of p_0 in $\mathbb{P}$ is compatible with some member of D.

Let $\mathbb{P}$ be a poset, λ an uncountable cardinal such that $\mathbb{P} \in H_\lambda$. Let $N \prec H_\lambda$ be countable with $\mathbb{P} \in N$. A condition $p \in \mathbb{P}$ is said to be <u>$(\mathbb{P},N)$-generic</u> iff, whenever $D \in N$ is predense in $\mathbb{P}$, then $D \cap N$ is predense below p in $\mathbb{P}$.

<u>Lemma 2.3</u>

Let $\mathbb{P}, \lambda, N$ be as above. A condition $p \in \mathbb{P}$ is $(\mathbb{P},N)$-generic iff, whenever $D \in N$ is <u>dense</u> in $\mathbb{P}$ then $D \cap N$ is predense below p in $\mathbb{P}$.

Proof: Since any dense set is predense, one direction is trivial. For the converse, suppose $p \in \mathbb{P}$ is such that whenever $D \in N$ is dense in $\mathbb{P}$, then $D \cap N$ is predense below p in $\mathbb{P}$. We show that p is $(\mathbb{P},N)$-generic.

Let $D \in N$ be pre-dense in $\mathbb{P}$. Set

$$E = \{q \in \mathbb{P} \mid (\exists d \in D)(q \leq d)\}.$$

Then E is dense in $\mathbb{P}$. Since $\mathbb{P}, D \in N \prec H_\lambda$, we have $E \in N$. So $E \cap N$ is predense below p in $\mathbb{P}$. Let $q \leq p$. Then q is compatible with some $r \in E \cap N$. Now,

$$H_\lambda \models (\exists d \in D)(r \leq d).$$

So, as $\mathbb{P}, D, r \in N \prec H_\lambda$, we can find a $d \in D \cap N$ such that $r \leq d$. Then q is compatible with d. Since $q \leq p$ was arbitrary, we are done.□

The following lemma might throw some light on the notion of $(\mathbb{P},N)$-genericity.

<u>Lemma 2.4</u>

Let $\mathbb{P}$ be a poset, λ a regular cardinal such that $\mathcal{P}(\mathbb{P}) \in H_\lambda$. If G is V-generic on $\mathbb{P}$, then G is H_λ-generic on $\mathbb{P}$ and $[H_\lambda]^{V[G]} = H_\lambda[G]$. Moreover, if $N \prec H_\lambda$ is countable with $\mathbb{P} \in N$ and if $p \in \mathbb{P}$ is $(\mathbb{P},N)$-generic, then $p \in G$ implies that $G \cap N$ is N-generic on $\mathbb{P} \cap N$.

Proof: Since $\mathcal{P}(\mathbb{P}) \in H_\lambda$, it is easily seen that G is H_λ-generic on $\mathbb{P}$, that λ is a regular cardinal in V[G], and that $[H_\lambda]^{V[G]} = H_\lambda[G]$. For the second part of the lemma, let $D \in N$ be such that $D \cap N$ is a dense subset of $\mathbb{P} \cap N$. Then, clearly,

$$N \models \text{"D is a dense subset of } \mathbb{P}\text{"}.$$

So as $N \prec H_\lambda$, we have

$$H_\lambda \models \text{"D is a dense subset of } \mathbb{P}\text{"}.$$

Hence D really is a dense subset of $\mathbb{P}$. Since $D \in N$, it follows that $D \cap N$ is predense below p in $\mathbb{P}$, i.e.

$$(\forall q \leq p)(\exists r \leq q)(\exists s \in D \cap N)(r \leq s).$$

Hence

$$p \Vdash_{\mathbb{P}} \text{"}(\exists r \in \mathring{G})(\exists s \in \widetilde{D \cap N})(r \leq s).\text{"}$$

Hence

$$p \Vdash_{\mathbb{P}} \text{"}(\widetilde{D \cap N}) \cap \mathring{G} \neq \phi\text{"}.$$

Since $p \in G$, it follows that in $V[G]$, $D \cap N \cap G \neq \phi$, and the result follows at once. □

We write $N \prec_{\mathbb{P}} H_\lambda$ if $N \prec H_\lambda$ is countable, $\mathbb{P} \in N$, and for every $p \in \mathbb{P} \cap N$ there is a $q \in \mathbb{P}$, $q \leq p$, such that q is $(\mathbb{P}, N)$-generic.

Lemma 2.5

The following are equivalent:

(i) $\mathbb{P}$ is proper;

(ii) There is a regular cardinal λ such that $\mathbb{P} \in H_\lambda$ and the set

$$\{N \in [H_\lambda]^{\aleph_0} \mid N \prec_{\mathbb{P}} H_\lambda\}$$

contains a club;

(iii) For all regular cardinals λ such that $\mathbb{P} \in H_\lambda$, the set

$$\{N \in [H_\lambda]^{\aleph_0} \mid N \prec_{\mathbb{P}} H_\lambda\}$$

contains a club.

Proof: (i)→(ii). Choose λ regular so that $\mathbb{P} \in H_\lambda$. Let

$$S = \{N \in [H_\lambda]^{\aleph_0} \mid \mathbb{P} \in N \prec H_\lambda \ \& \ N \not\prec_{\mathbb{P}} H_\lambda\}.$$

Now, the set

$$\{N \in [H_\lambda]^{\aleph_0} \mid \mathbb{P} \in N \prec H_\lambda\}$$

is club, so if (ii) were to fail, S would be stationary. Let us suppose this were the case. Define $\rho : S \to H_\lambda$ by letting $\rho(N)$ be some member of $\mathbb{P} \cap N$, no extension of which is $(\mathbb{P},N)$-generic. By lemma 1.4, there is a stationary set $\bar{S} \subseteq S$ such that $\rho(N) = p$ for all $N \in \bar{S}$, where $p \in \mathbb{P}$ is fixed. Let G be a V-generic subset of $\mathbb{P}$ containing p. In V[G], define

$$C = \{N \in [H_\lambda^V]^{\aleph_0} \mid N \prec H_\lambda^V \text{ \& for all } D \in N \text{ which are predense in } \mathbb{P} \text{ there is a } q \in G \text{ such that } q \in D \cap N\}.$$

Clearly, C is club in $[H_\lambda^V]^{\aleph_0}$ (in V[G]). So as $\mathbb{P}$ is proper there is an $N \in \bar{S} \cap C$. Notice that $N \in V$ and $N \prec H_\lambda^V$. Choose $q \in G$, $q \le p$, so that $q \Vdash_{\mathbb{P}}$ "$\check{N} \in \mathring{C}$".

We work in V from now on. We have:

For all pre-dense $D \in N$, $q \Vdash_{\mathbb{P}}$ "$(\exists r \in \mathring{G})(r \in \check{D} \cap \check{N})$".

So, it must be the case that:

For all pre-dense $D \in M$ there is a $q' \le q$ such that $q' \le r$ for some $r \in D \cap N$. Hence q is $(\mathbb{P},N)$-generic. But $q \le p = \rho(N)$, so we have a contradiction. This proves (ii). (In fact we have proved (iii), but we still need to know that (ii) $\to$ (iii).)

(ii)$\to$ (iii). Let λ be as in (ii). Let μ be regular with $\mathbb{P} \in H_\mu$. We show that

$$C = \{N \in [H_\mu]^{\aleph_0} \mid N \prec_{\mathbb{P}} H_\mu\}$$

contains a club set. If $\mu = \lambda$, there is nothing to prove. There are two remaining cases.

Suppose that $\mu < \lambda$. Let

$$C_1 = \{N \cap H_\mu \mid N \prec_{\mathbb{P}} H_\lambda\}.$$

By lemma 1.8(i), C_1 contains a club in $[H_\mu]^{\aleph_0}$. Let

$$C_2 = \{N \in [H_\mu]^{\aleph_0} \mid N \prec H_\mu\}.$$

Then C_2 is club in $[H_\mu]^{\aleph_0}$. Hence $C_1 \cap C_2$ contains a club in $[H_\mu]^{\aleph_0}$. But clearly,

$$C_1 \cap C_2 \subseteq C,$$

so we are done in this case.

Now suppose that $\mu > \lambda$. Let

$$C_1 = \{N \in [H_\mu]^{\aleph_0} \mid N \cap H_\lambda \prec_{\mathbb{P}} H_\lambda\}.$$

By 1.8(ii), C_1 contains a club in $[H_\mu]^{\aleph_0}$. Let C_2 be as above. Then $C_1 \cap C_2$ contains a club in $[H_\mu]^{\aleph_0}$ and $C_1 \cap C_2 \subseteq C$, so again we are done.

(iii)→(i). Let $S \subseteq [\lambda]^{\aleph_0}$ be stationary. We show that

$$\Vdash_{\mathbb{P}} \text{"}\check{S} \text{ is stationary"}.$$

Let $\mathring{C} \in V^{(\mathbb{P})}$ be such that

$$\Vdash_{\mathbb{P}} \text{"}\mathring{C} \text{ is club in } [\check{\lambda}]^{\aleph_0}\text{"}.$$

We prove that

$$\Vdash_{\mathbb{P}} \text{"}\mathring{C} \cap \check{S} \neq \phi\text{"}.$$

Pick $\mathring{f} \in V^{(\mathbb{P})}$ so that

$$\Vdash_{\mathbb{P}} \text{"}\mathring{f} : [\check{\lambda}]^{<\aleph_0} \to \check{\lambda} \text{ concentrates on } \mathring{C}\text{"}.$$

If we can find an $X \in S$ such that

$$\Vdash_{\mathbb{P}} \text{"}\mathring{f}\text{"}[\check{X}]^{<\aleph_0} \subseteq \check{X}\text{"}$$

we shall be done.

Choose μ regular so that $\mathcal{P}(\mathbb{P}), \mathring{f}, \lambda \in H_\mu$. By assumption, the set

$$E = \{N \in [H_\mu]^{\aleph_0} \mid \lambda, \mathring{f} \in N \prec_{\mathbb{P}} H_\mu\}$$

contains a club.

Let $p \in \mathbb{P}$ be given, and set

$$E_p = \{N \in E \mid p \in N\}.$$

Then E_p contains a club, so by 1.8(i) the set

$$E_p^\lambda = \{N \cap \lambda \mid N \in E_p\}$$

contains a club. Thus $E_p^\lambda \cap S \neq \phi$. Choose $N \in E_p$ so that $N \cap \lambda \in S$. We finish the proof by showing that if $X = N \cap \lambda$, then for some $q \le p$,

$$q \Vdash_{\mathbb{P}} \text{"} \mathring{f}\text{"}[\check{X}]^{<\aleph_0} \subseteq \check{X}\text{"}.$$

Pick $q \le p$ to be $(\mathbb{P},N)$-generic. Let $\sigma \in [X]^{<\aleph_0}$ and set

$$D = \{r \in \mathbb{P} \mid (\exists\alpha<\lambda)(r \Vdash_{\mathbb{P}} \text{"}\mathring{f}(\check{\sigma}) = \check{\alpha}\text{"})\}$$

Then D is dense in $\mathbb{P}$. Moreover, since $\mathbb{P},\mathring{f},\lambda,\sigma \in N \prec H_\mu$, $D \in N$. Thus $D \cap N$ is pre-dense below q in $\mathbb{P}$. Let $q' \le q$ be such that

$$q' \Vdash_{\mathbb{P}} \text{"}\mathring{f}(\check{\sigma}) = \check{\alpha}\text{"}.$$

By extending q' if necessary, we may assume that $q' \le r$ for some $r \in D \cap N$. Since $r \in D$ and $q' \le r$ we have

$$r \Vdash_{\mathbb{P}} \text{"}\mathring{f}(\check{\sigma}) = \check{\alpha}\text{"}.$$

But $r, \mathbb{P}, \mathring{f}, \check{\sigma} \in N \prec H_\mu$. Hence $\alpha \in N$. It follows that

$$q \Vdash_{\mathbb{P}} \text{"}\mathring{f}(\check{\sigma}) \in \check{X}\text{"},$$

so we are done. □

We give a model-theoretic characterisation of properness. Suppose $\mathbb{P} \in H_\lambda$. We say that a structure

$$M = \langle H_\lambda, \in, \mathbb{P}, \ldots\rangle$$

is <u>$\mathbb{P}$-good</u> iff

$$N \prec M \ \& \ |N| = \aleph_0 \text{ implies } N \prec_{\mathbb{P}} H_\lambda.$$

<u>Lemma 2.6</u>

The following are equivalent:

(i) $\mathbb{P}$ is proper.

(ii) If λ is a regular cardinal such that $\mathbb{P} \in H_\lambda$, and if $M = \langle H_\lambda, \in, \ldots \rangle$, then there is an expansion M' of M which is $\mathbb{P}$-good.

(iii) There is a $\mathbb{P}$-good structure.

Proof: (i)$\to$(ii). By lemma 2.5 there is a club set $C \subseteq [H_\lambda]^{\aleph_0}$ such that

$$N \in C \to N \prec_{\mathbb{P}} H_\lambda.$$

By lemma 1.7 there is an expansion M' of M such that

$$N \prec M' \ \& \ |N| = \aleph_0 \to N \in C.$$

(ii)$\to$(iii). Trivial.

(iii)$\to$(i). By lemma 1.7, the countable elementary submodels of a $\mathbb{P}$-good structure form a club set. So by lemma 2.5, $\mathbb{P}$ is proper. $\square$

§3. PROPER ITERATIONS

The following result is an immediate consequence of the definition of properness:

Theorem 3.1

If $\mathbb{P}$ is a proper poset and $\mathring{Q} \in V^{(\mathbb{P})}$ is such that $\|\mathring{Q}$ is a proper poset$\|^{\mathbb{P}} = \mathbb{1}$, then $\mathbb{P} \otimes \mathring{Q}$ is a proper poset. $\square$

We shall show that any countable support iteration of proper posets is proper. We begin by considering the special case of an iteration of length ω. The general case is basically just a generalisation of this case, so this approach will not result in any great duplication of effort.

Theorem 3.2

Let $\langle \mathbb{P}_\nu \mid \nu \leq \omega \rangle$ be a countable support iteration of posets such that $\mathbb{P}_0$ is proper and for each $n < \omega$ there is a $\mathbb{Q}_{n+1} \in V^{(\mathbb{P}_n)}$ such that

$$\| \mathbb{Q}_{n+1} \text{ is a proper poset } \|^{\mathbb{P}_n} = \mathbf{1}$$

and $\mathbb{P}_{n+1} = \mathbb{P}_n \otimes \mathbb{Q}_{n+1}$. Then $\mathbb{P}_\omega$ is proper. (By Theorem 3.1, each $\mathbb{P}_n$, $n < \omega$, is proper.)

Proof: We use Lemma 2.6 in order to prove that $\mathbb{P}_\omega$ is proper. Pick λ a regular cardinal such that $\mathcal{P}(\mathbb{P}_\omega) \in H_\lambda$. We construct a $\mathbb{P}_\omega$-good structure

$$M = \langle H_\lambda, \in, \mathbb{P}_\omega, \ldots \rangle \quad .$$

We shall assume that for each $\nu \leq \omega$, $\mathbb{P}_\nu$ consists of sequences $q = \langle q(n) \mid n < 1 + \nu \rangle$ such that $q(0) \in \mathbb{Q}_0$ (where we set $\mathbb{Q}_0 = \mathbb{P}_0$ for convenience) and for each $n < \nu$, $q(n+1)$ is an element of $V^{(\mathbb{P}_n)}$ such that $q \restriction n+1 \Vdash_{\mathbb{P}_n}$ "$q(n+1) \in \mathbb{Q}_{n+1}$", with the ordering

$$q' \leq_\nu q \text{ iff } q'(0) \leq_0 q(0) \ \& \ (\forall n < \nu)[q' \restriction n+1 \Vdash_{\mathbb{P}_n} \text{"} q'(n+1) \leq_{n+1} q(n+1) \text{"}].$$

We define sequences $\langle M_n \mid n < \omega \rangle$ and $\langle f_{n+1} \mid n < \omega \rangle$ as follows. Let $\leq^*$ be a well-ordering of H_λ, and let $M_0 = \langle H_\lambda, \in, \mathbb{P}_\omega, \mathbb{Q}_0, \leq^*, \ldots \rangle$ (of countable length) be any fixed $\mathbb{Q}_0$-good structure. For each $n < \omega$, let M_{n+1} be an element of $V^{(\mathbb{P}_n)}$ such that

$$\| M_{n+1} = \langle H_\lambda, \in, \mathbb{P}_\omega, \mathbb{Q}_{n+1}, G_n, \ldots \rangle \text{ is } \mathbb{Q}_{n+1}\text{-good} \|^{\mathbb{P}_n} = \mathbf{1} \ .$$

For each $n < \omega$, f_{n+1} is an element of $V^{(\mathbb{P}_n)}$ such that

$$||f_{n+1} : [H_\lambda]^{<\aleph_0} \to H_\lambda \text{ is an } M_{n+1}\text{-skolem function}||^{\mathbb{P}_n} = \mathbb{1}$$

(in the sense of lemma 1.6).

Now, by lemma 2.4, if G_n is V-generic on $\mathbb{P}_n$, then $[H_\lambda]^{V[G_n]} = H_\lambda{}^V[G_n]$. Hence, every element of $H_\lambda{}^{V[G_n]}$ is denoted by a boolean $\mathbb{P}_n$-name which lies in H_λ^V. [But notice that elements of $H_\lambda{}^{V[G_n]}$ may also have names which lie outside of $H_\lambda{}^V$. Indeed, the "obvious" boolean name of many elements will involve a reference to λ, and no such name will lie in $H_\lambda{}^V$, of course. It will later be observed that this is the motivation behind our following definitions.]

For each $n < \omega$, define a function

$$I_n : [H_\lambda]^{<\aleph_0} \to H_\lambda$$

such that whenever $\sigma = \{x_1,\ldots,x_n\}$ is a set of boolean $\mathbb{P}_n$-names (i.e. elements of $V^{(\mathbb{P}_n)}$) such that $||x_i \in H_\lambda||^{\mathbb{P}_n} = \mathbb{1}$, then $I_n(\sigma)$ is a maximal pairwise incompatible subset of $\mathbb{P}_n$ with the property that if $p \in I_n(\sigma)$, then for some boolean $\mathbb{P}_n$-name x in H_λ, $p \Vdash_{\mathbb{P}_n}$ "$f_{n+1}(\sigma) = x$". And define a further function

$$h_n : \mathbb{P}_n \times [H_\lambda]^{<\aleph_0} \to H_\lambda$$

by letting $h_n(p,\sigma) = x$ for some $\mathbb{P}_n$-name x in H_λ such that $p \Vdash_{\mathbb{P}_n}$ "$f_{n+1}(\sigma) = x$". The behaviour of both I_n and h_n in any case not covered above is irrelevant to our purposes.

Let M be the structure

$$M = \langle M_o, (I_n)_{n<\omega}, (h_n)_{n<\omega} \rangle.$$

We show that M is $\mathbb{P}_\omega$-good.

Let $N \prec M$ be countable. We must show that $N \prec_{\mathbb{P}_\omega} H_\lambda$. So, given some $p \in \mathbb{P}_\omega \cap N$ we must find a $q \in \mathbb{P}_\omega$, $q \le p$, such that whenever $D \in N$ is dense in $\mathbb{P}_\omega$, then $D \cap N$ is pre-dense below q in $\mathbb{P}_\omega$.

Fix some enumeration $\langle D_n | n<\omega \rangle$ of all dense subsets of $\mathbb{P}_\omega$ lying in N. The idea is now to obtain q by defining $q(n), n<\omega$, inductively, at stage n ensuring that $D_n \cap N$ will be pre-dense below q in $\mathbb{P}_\omega$.

Since D_0 is dense in $\mathbb{P}_\omega$,

$$M \models (\exists p_0 \in D_0)(p_0 \le p).$$

But $\mathbb{P}_\omega, D_0, p \in N \prec M$, so we can find a $p_0 \in D_0 \cap N$ such that $p_0 \le p$. We shall ensure that we choose q so that $q \le p_0$ in $\mathbb{P}_\omega$, which will take care of the dense set D_0. Dealing with the other dense sets will be a little more tricky.

Now, M_0 is $\mathbb{Q}_0$-good and $p_0(0) \in N \prec M_0$, so there is a $(\mathbb{Q}_0, N)$-generic $q(0) \in \mathbb{Q}_0$ such that $q(0) \le p_0(0)$. Before we define $q(1)$, we need an auxiliary notion and a couple of technical results.

For each $n < \omega$, if G_n is V-generic on $\mathbb{P}_n$, then in $V[G_n]$ we can consider the set, N_{n+1}, of all elements of $H_\lambda^{V[G_n]}$ which have a boolean $\mathbb{P}_n$-name inside N. We let $\mathring{N}_{n+1}$ be a boolean $\mathbb{P}_n$-name for the set N_{n+1}. The proofs of the following two results are deferred until later:

<u>Claim 1</u>. If $q \in \mathbb{P}_n$ is $(\mathbb{P}_n, N)$-generic, then

$$q \Vdash_{\mathbb{P}_n} \text{"}\mathring{N}_{n+1} \prec M_{n+1}\text{"}. \qquad \blacktriangle$$

<u>Claim 2</u>. If $q \in \mathbb{P}_n$ is $(\mathbb{P}_n, N)$-generic, and if $q' \in V^{(\mathbb{P}_n)}$ is such that

$$q \Vdash_{\mathbb{P}_n} "q' \in \mathbb{Q}_{n+1} \text{ is } (\mathbb{Q}_{n+1}, \mathring{N}_{n+1})\text{-generic}",$$

then $q \frown \langle q' \rangle$ is $(\mathbb{P}_{n+1}, N)$-generic. △

Now, $q(1)$ will be an element of $V^{(\mathbb{P}_0)}$ such that $q(0) \Vdash_{\mathbb{P}_0} "q(1) \in \mathbb{Q}_1"$. So, in order to define $q(1)$ it is permissible to proceed as follows. Pick any V-generic set G_0 on $\mathbb{P}_0$ containing $q(0)$. Inside $V[G_0]$, construct "$q(1)$" as required. Then back in V apply the maximum principle to make the actual choice of $q(1)$ as a name for the element constructed in $V[G_0]$. (Of course, this could all be done without any mention of generic sets as such, but for clarity we shall proceed as indicated.)

So pick G_0 V-generic on $\mathbb{P}_0$ with $q(0) \in G_0$ now. We work in $V[G_0]$ until further notice. Let

$$E_1 = \{r \in \mathbb{P}_0 \mid (\exists s \leq p_0)[s \in D_1 \ \& \ s \restriction 1 = r]\}.$$

Clearly, E_1 is dense below $p_0 \restriction 1$ in $\mathbb{P}_0$. Moreover, $E_1 \in H_\lambda^V$. Indeed, since we have $\mathbb{P}_0, p_0, D_1 \in N \prec H_\lambda^V$, we have $E_1 \in N$. Let $A_1 \subseteq E_1$, $A_1 \in V$, be the $<^*$-least maximal antichain below $p_0 \restriction 1$ in $\mathbb{P}_0$. Thus $A_1 \in N$. For each $r \in A_1$, let r^* be the $<^*$-least $r^* \in D_1$ such that $r^* \leq p_0$ and $r^* \restriction 1 = r$. Notice that $r \in N \to r^* \in N$.

<u>Claim 3</u>. $G_0 \cap (A_1 \cap N) \neq \phi$

Proof: We have : $A_1 \in N$ is pre-dense below $p_0(0)$ in $\mathbb{P}_0$ and $q(0) \leq p_0(0)$ and $q(0)$ is $(\mathbb{P}_0, N)$-generic. Hence every extension of $q(0)$ in $\mathbb{P}_0$ is compatible with some member of $A_1 \cap N$. But G_0 is V-generic on $\mathbb{P}_0$ and contains $q(0)$. Hence $G_0 \cap (A_1 \cap N) \neq \phi$.△

Let $r \in G_0 \cap (A_1 \cap N)$. Clearly, r is uniquely determined here. Let $p_1 = r^*$ for this r. We have

$$p_1 \leq p_0, \ p_1 \in D_1 \cap N, \ p_1 \restriction 1 \in G_0.$$

(But notice that although $p_1 \in V$, the actual choice of p_1 depends upon G_0. Different sets G_0 may give rise to different conditions p_1.)

By claim 1, since $q(0) \in G_0$, we have $N_1 \prec M_1$. Moreover, $p_1 \in N$, so $p_1(1) \in N_1$. So, as M_1 is $\mathbb{Q}_1$-good there is a $q(1) \in \mathbb{Q}_1$, $q(1) \leq p_1(1)$, $q(1)$ $(\mathbb{Q}_1, N_1)$-generic.

Returning to V now, we pick $q(1) \in V^{(\mathbb{P}_0)}$ a name of the element q(1) just defined. (There is no sense in trying to distinguish these two "q(1)-s".) Then $(q(0),q(1)) \in \mathbb{P}_1$ and, by claim 2, $(q(0),q(1))$ is $(\mathbb{P}_1,N)$-generic.

To choose q(2) now, we let G_1 be V-generic on $\mathbb{P}_1$ with $q\restriction 2 \in G_1$ and work inside $V[G_1]$. Let $G_0 = \{r \restriction 1 \mid r \in G_0\}$. Then G_0 is V-generic on $\mathbb{P}_0$ and $q(0) \in G_0$. Let p_1 be the condition chosen as above for this G_0. Set

$$E_2 = \{r \in \mathbb{P}_1 \mid (\exists s \leq p_1)[s \in D_2 \ \& \ s\restriction 2 = r)\}.$$

E_2 is dense below $p_1\restriction 2$ in $\mathbb{P}_1$ and $E_2 \in N$. Let $A_2 \subseteq E_2$ be the $<^*$-least maximal antichain below $p_1\restriction 2$ in $\mathbb{P}_1$. Thus $A_2 \in N$. For each $r \in A_2$, let r^* be the $<^*$-least $r^* \in D_2$ such that $r^* \leq p_1$ and $r^*\restriction 2 = r$. So $r \in N \to r^* \in N$.

<u>Claim 4</u> : $G_1 \cap (A_2 \cap N) \neq \phi$.

Proof: We have $p_1 \restriction 1 \in G_0$. Hence $s = p_1(0) \wedge q(0) \in G_0$. Moreover, since $s\wedge(q\restriction 2) \leq q\restriction 2$, $s\wedge(q\restriction 2)$ is $(\mathbb{P}_1,N)$-generic. But $A_2 \in N$

is pre-dense below $p_1 \restriction 2$ in $\mathbb{P}_1$ and $s \wedge (q \restriction 2) \leq p_1 \restriction 2$, so every extension of $s \wedge (q \restriction 2)$ in $\mathbb{P}_1$ is compatible with some element of $A_2 \cap N$. Since $s \wedge (q \restriction 2) \in G_1$, the claim follows immediately. ▲

Let $r \in G_1 \cap (A_2 \cap N)$ and set $p_2 = r^*$ for this unique r. Then

$$p_2 \leq p_1, \quad p_2 \in D_2 \cap N, \quad p_2 \restriction 2 \in G_1.$$

By claim 1 now, since $q \restriction 2 \in G_1$, we have $N_2 \prec M_2$. Moreover $p_2 \in N$, so $p_2(2) \in N_2$. Thus as M_2 is $\mathbb{Q}_2$-good there is a $q(2) \in \mathbb{Q}_2, q(2) \leq p_2(2)$, $q(2)$ $(\mathbb{Q}_2, N_2)$-generic.

This defines $q(2) \in V^{(\mathbb{P}_1)}$ so that $q \restriction 3$ is (by claim 2) $(\mathbb{P}_2, N)$-generic. We continue in this fashion to define $q(3), q(4), \ldots, q(n), \ldots (n<\omega)$. In general, $q(n+1)$ is chosen in $V[G_n]$ so that $q(n+1)$ is $(\mathbb{Q}_{n+1}, N_{n+1})$-generic, and $q(n+1) \leq p_{n+1}(n+1)$, where $p_{n+1} \leq p_n$, $p_{n+1} \in D_{n+1} \cap N$, and $p_{n+1} \restriction n+1 \in G_n$ (the actual choice of p_{n+1} being dependent upon G_n). [Notice that although Claim 3 was a special case (since $q(0) \leq p_0(0)$), Claim 4 generalises to any n+1. In order to show that $G_{n+1} \cap (A_{n+2} \cap N) \neq \phi$, where $n \geq 0$, we utilise the facts that $(p_{n+1} \restriction n+1) \wedge (q \restriction n+1) \in G_n$ and, since $(p_{n+1} \restriction n+1) \wedge (q \restriction n+2) \leq q \restriction n+2$, $(p_{n+1} \restriction n+1) \wedge (q \restriction n+2)$ is $(\mathbb{P}_{n+1}, N)$-generic.]

Having defined $q = \langle q(n) \mid n<\omega \rangle$ now, we must prove that q is $(\mathbb{P}_\omega, N)$-generic.

So, given $q' \leq q$ we must show that for any $n<\omega$, q' is compatible with some member of $D_n \cap N$. For $n = 0$ we have $q' \leq q \leq p_0 \in D_0 \cap N$, so there is nothing to prove. We deal with all of the remaining cases in V[G], where G is any V-generic subset of $\mathbb{P}_\omega$ containing q'. This clearly causes no loss of generality, as we are dealing entirely with members of V. The advantage is that, working in V[G] from now on, if we set $G_n = \{r\restriction n+1 \mid r \in G\}$, then G_n is a V-generic subset of $\mathbb{P}_n$ containing $q\restriction n+1$, for each $n<\omega$, and there is a well-defined sequence $p_0 \geq p_1 \geq p_2 \geq \ldots\ p_n \geq \ldots (n<\omega)$, as described above, and moreover since $q(n) \leq p_n(n)$ for all $n<\omega$, we have, for each $n<\omega, q(m) \leq p_n(m)$ for all $m \geq n$. Let us look at the case n=1 now. Now, $A_1 \in N$ is pre-dense below $p_0\restriction 1$ in $\mathbb{P}_0$ and q(0) is $(\mathbb{P}_0,N)$-generic and $q'(0) \leq q(0) \leq p_0\restriction 1$, so there is an $s \in G_0$ such that $s \leq q'(0)$ and $s \leq r$ for some $r \in A_1 \cap N$. But then $p_1 = r^*$. Hence, since we have $q'(m) \leq q(m) \leq p_1(m)$ for all $m > 0$, we conclude that $s \wedge q' \leq p_1$. But $p_1 \in D_1 \cap N$, so we are done in this case. All other cases are handled in a similar fashion. For instance, looking at n = 2, we know that $A_2 \in N$ is pre-dense below $p_1\restriction 2$ in $\mathbb{P}_1$ and $q\restriction 2$ is $(\mathbb{P}_1,N)$-generic and (since $q'\restriction 2 \in G_1$) $(p_1\restriction 2)\wedge(q'\restriction 2) \leq (p_1\restriction 2),(q'\restriction 2)$ and $(p_1\restriction 2)\wedge(q'\restriction 2) \in G_1$, so there is an extension, s, of $(p_1\restriction 2)\wedge(q'\restriction 2)$ in G_1 such that $s \leq r$ for some $r \in A_2 \cap N$. Then $r \in G_1$, so $r^* = p_2$, and $s \wedge q' \leq p_2 \in D_2 \cap N$, so we are again done. (The case n = 2 is typical of any case $n \geq 2$, of course.) That proves that q is $(\mathbb{P}_\omega,N)$-generic.

We are left with the proof of claims 1 and 2. We begin with claim 1. So let $q \in \mathbb{P}_n$ be $(\mathbb{P}_n,N)$-generic. We want to show that

$$q \Vdash_{\mathbb{P}_n} "\mathring{N}_{n+1} \prec M_{n+1}".$$

It suffices to show that

$$q \Vdash_{\mathbb{P}_n} "f_{n+1} "[\mathring{N}_{n+1}]^{<\aleph_0} \subseteq \mathring{N}_{n+1}".$$

So, given $\mathbb{P}_n$-names $x_1,\ldots,x_m \in N$ such that $\|x_i \in H_\lambda\|^{\mathbb{P}_n} = \mathbf{1}$ for each i, we must show that $q \Vdash_{\mathbb{P}_n} "f_{n+1}(\{x_1,\ldots,x_m\}) \in \mathring{N}_{n+1}"$. Now, since $x_1,\ldots,x_m \in N$ and $N \prec M$, we have $I_n(\{x_1,\ldots,x_m\}) \in N$. Moreover, $I_n(\{x_1,\ldots,x_m\})$ is a maximal pairwise incompatible subset of $\mathbb{P}_n$. So, if we are given any $q' \leq q$, then by the $(\mathbb{P}_n,N)$-genericity of q there is a $q'' \leq q'$ such that $q'' \leq r$ for some $r \in I_n(\{x_1,\ldots,x_m\})$. Since $r \in I_n(\{x_1,\ldots,x_m\})$, $x = h_n(r_1 \{x_1,\ldots,x_m\})$ is defined. We have

$$r \Vdash_{\mathbb{P}_n} "f_{n+1}(\{x_1,\ldots,x_m\}) = x".$$

So as $q'' \leq r$,

$$q'' \Vdash_{\mathbb{P}_n} "f_{n+1}(\{x_1,\ldots,x_m\}) = x".$$

But $r, x_1,\ldots x_m \in N \prec M$. Hence $x \in N$, and we conclude that

$$q'' \Vdash_{\mathbb{P}_n} "f_{n+1}(\{x_1,\ldots,x_m\}) \in \mathring{N}_{n+1}".$$

Since $q'' \leq q'$ and $q' \leq q$ was arbitrary, this completes the proof of claim 1.

We turn to claim 2. Let $q \in \mathbb{P}_n$ be $(\mathbb{P}_n,N)$-generic, and let $q' \in V^{(\mathbb{P}_n)}$ be such that $q \Vdash_{\mathbb{P}_n} "q' \in \mathbb{Q}_{n+1}$ is $(\mathbb{Q}_{n+1},\mathring{N}_{n+1})$-generic". We show that $q^\frown\langle q'\rangle$ is $(\mathbb{P}_{n+1},N)$-generic.

Let $D \in N$ be a dense subset of $\mathbb{P}_{n+1}$. We show that if $q_1{}^\frown\langle q_1'\rangle \leq q^\frown\langle q'\rangle$, then $q_1{}^\frown\langle q_1'\rangle$ is compatible with some member of $D \cap N$.

Let $D_2 \in V^{(\mathbb{P}_n)}$ be such that whenever G_n is V-generic on $\mathbb{P}_n$, then in $V[G_n]$,

$$D_2 = \{r \in \mathbb{Q}_{n+1} \mid (\exists s \in G_n)[s^\frown\langle r\rangle \in D]\}.$$

The boolean object D_2 is definable from $\mathbb{P}_{n+1}, \mathring{G}_n, D$ in the H_λ, so as $\mathbb{P}_{n+1}, \mathring{G}_n, D \in N \prec H_\lambda$, we have $D_2 \in N$. We show that

$$\|D_2 \text{ is dense in } \mathbb{Q}_{n+1}\|^{\mathbb{P}_n} = 1$$

To verify this we argue in $V[G_n]$. Let $r \in \mathbb{Q}_{n+1}$ be given. Pick $s \in G_n$ so that $s \Vdash_{\mathbb{P}_n}$ "$r \in \mathbb{Q}_{n+1}$". Then $s^\frown\langle r\rangle \in \mathbb{P}_{n+1}$. Suppose $s' \leq_{\mathbb{P}_n} s$. Then $s'^\frown\langle r\rangle \in \mathbb{P}_{n+1}$ so there is an $s''^\frown\langle r''\rangle \in D$ such that $s''^\frown\langle r''\rangle \leq_{\mathbb{P}_{n+1}} s'^\frown\langle r\rangle$. Clearly

$$s'' \Vdash_{\mathbb{P}_n} \text{"} r'' \leq r \;\&\; r'' \in D_2\text{"}.$$

Hence the set

$$\{s'' \in \mathbb{P}_n \mid s'' \Vdash_{\mathbb{P}_n} \text{"}(\exists r'' \leq r)(r'' \in D_2)\text{"}\}$$

is dense below s in $\mathbb{P}_n$. But $s \in G_n$. Hence there is an $r'' \leq r$, $r'' \in D_2$, as required.

Now, since $q_1 \leq q$, we have

$$q_1 \Vdash_{\mathbb{P}_n} \text{"} q' \in \mathbb{Q}_{n+1} \text{ is } (\mathbb{Q}_{n+1}, \mathring{N}_{n+1})\text{-generic"}.$$

But $q_1 \Vdash_{\mathbb{P}_n}$ "$q_1' \leq q'$". Hence as $D_2 \in N$,

$$q_1 \Vdash_{\mathbb{P}_n} \text{"}(\exists r \in D_2 \cap \mathring{N}_{n+1})(r \sim q_1')\text{."}$$

So we can find a $q_2 \leq q_1$ and an $r \in V^{(\mathbb{P}_n)} \cap N$ such that

$$q_2 \Vdash_{\mathbb{P}_n} \text{"} r \in D_2 \;\&\; r \sim q_1'\text{"}.$$

Let

$$D_1 = \{s \in \mathbb{P}_n \mid (s \Vdash_{\mathbb{P}_n} \text{"} r \notin D_2\text{"}) \text{ or } (\exists s' \geq s)[s'^\frown\langle r\rangle \in D]\}.$$

We claim that D_1 is dense in $\mathbb{P}_n$. For suppose $s \in \mathbb{P}_n$. If there is an $s' \leq s$ such that $s' \Vdash_{\mathbb{P}_n}$ "$r \notin D_2$", then $s' \in D_1$ and we are done. Otherwise, $s \Vdash_{\mathbb{P}_n}$ "$r \in D_2$", so by definition of D_2, $s \Vdash_{\mathbb{P}_n}$ "$(\exists s' \in \mathring{G}_n)[s'^\frown\langle r\rangle \in D]$". Hence there is an $s'' \leq s$ and an $s' \in \mathbb{P}_n$ such that $s'' \Vdash_{\mathbb{P}_n}$ "$s' \in \mathring{G}_n$ & $s'^\frown\langle r\rangle \in D$". Since $s'' \Vdash_{\mathbb{P}_n}$ "$s' \in \mathring{G}_n$", we must have $s'' \leq s'$. And if $s'' \Vdash_{\mathbb{P}_n}$ "$s'^\frown\langle r\rangle \in D$", we must in fact have $s'^\frown\langle r\rangle \in D$. Hence $s'' \in D_1$. Since $s'' \leq s$, we are again done.

Now, D_1 is definable from $\mathbb{P}_{n+1}, D_2, r$ in H_λ and $\mathbb{P}_{n+1}, D_2, r \in N \prec H_\lambda$. Hence $D_1 \in N$. So, as $q_2 \leq q_1 \leq q$ there is an $s \in D_1 \cap N$ such that $q_2 \sim s$. Now, $q_2 \Vdash_{\mathbb{P}_n}$ "$r \in D_2$", so $s \nVdash_{\mathbb{P}_n}$ "$r \notin D_2$". So as $s \in D_1$ there must be some $s' \geq s$ such that $s'^\frown\langle r\rangle \in D$. Thus

$$H_\lambda \models (\exists s' \geq s)[s'^\frown\langle r\rangle \in D].$$

But $s, r, D, \mathbb{P}_{n+1} \in N \prec H_\lambda$. Hence we can find an $s' \in N$ such that $s' \geq s$ and $s'^\frown\langle r\rangle \in D$.

Now, $s' \in N$ and $r \in N$, so $s'^\frown\langle r\rangle \in D \cap N$. We know that $q_2 \sim s$, so as $s' \geq s$ we have $q_2 \sim s'$. Pick $u \leq q_2, s'$. Since $u \leq q_2$, $u \Vdash_{\mathbb{P}_n}$ "$r \sim q_1'$". So we can find a $v \in V^{(\mathbb{P}_n)}$ such that $u \Vdash_{\mathbb{P}_n}$ "$v \leq r, q_1'$". But $q_2 \leq q_1$, so we have $u^\frown\langle v\rangle \leq q_1^\frown\langle q_1'\rangle$, $s'^\frown\langle r\rangle$. Thus $q_1^\frown\langle q_1'\rangle$ is compatible with $s'^\frown\langle r\rangle$, which is in $D \cap N$, and we are done. □

We are now ready to prove the general iteration lemma. Since this is basically just a generalisation of the above proof, we shall simply give an outline of the alterations required.

<u>Theorem 3.3</u>

Let $\langle \mathbb{P}_\nu \mid \nu \leq \lambda\rangle$ be a countable support iteration of posets such that $\mathbb{P}_0$ is proper and for each $\nu < \lambda$ there is a

$\mathbb{Q}_{\nu+1} \in V^{(\mathbb{P}_\nu)}$ such that $\|\mathbb{Q}_{\nu+1}$ is a proper poset$\|^{\mathbb{P}_\nu} = \mathbb{1}$ and $\mathbb{P}_{\nu+1} = \mathbb{P} \otimes \mathbb{Q}_{\nu+1}$. Then $\mathbb{P}_\nu$ is proper for all $\nu \leq \lambda$.

Proof: By induction on ν. For $\nu = 0$ there is nothing to prove, and successor steps follow directly from Theorem 3.1. Suppose now that ν is a limit ordinal and the result holds for all $\tau < \nu$.

For $\alpha<\beta<\nu$, let $\mathbb{Q}_\alpha^\beta \in V^{(\mathbb{P}_\alpha)}$ be that $V^{(\mathbb{P}_\alpha)}$-poset such that $\mathbb{P}_\beta = \mathbb{P}_\alpha \otimes \mathbb{Q}_\alpha^\beta$. By the induction hypothesis, $\|\mathbb{Q}_\alpha^\beta$ is proper$\|^{\mathbb{P}_\alpha} = \mathbb{1}$. We fix a system of structures $\langle M_\alpha^\beta \mid \alpha < \beta < \nu\rangle$ such as in Theorem 3.2 so that M_0^β is $\mathbb{P}_\beta$-good for all $\beta<\nu$, and $\|M_\alpha^\beta$ is $\mathbb{Q}_\alpha^\beta$-good$\|^{\mathbb{P}_\alpha} = \mathbb{1}$ for all $\alpha<\beta<\nu$. Define $\langle f_\alpha^\beta \mid \alpha<\beta<\nu\rangle$ as before, and then define a structure

$$M = \langle H_\lambda, \in, \mathbb{P}_\nu, \leq^*, \ldots\rangle$$

again much as before in order to incorporate the necessary (forcing) information about the functions f_α^β. In order to ensure that the language of M is countable, however, we must include the indices α,β as <u>variables</u> rather than regarding them as metamathematical indices. So, instead of including functions $I_\alpha^\beta(\sigma)$, $h_\alpha^\beta(p,\sigma)$, $\alpha<\beta<\nu$, in M (to play the same role as their counterparts in Theorem 3.2), we have just the two functions $I(\alpha,\beta,\sigma)$ and $h(\alpha,\beta,p,\sigma)$. To show that M is $\mathbb{P}_\nu$-good, we commence as before. Let $N \prec M$ be countable. We want to prove that $N \prec_{\mathbb{P}_\nu} M$.

Let $\gamma = \sup(N\cap\nu)$, and let $\langle\gamma(n) \mid n<\omega\rangle$ be a strictly increasing sequence from $N\cap\gamma$ which is cofinal in γ. The idea is to proceed along the sequence $\langle\mathbb{P}_{\gamma(n)} \mid n<\omega\rangle$ (via the boolean posets

$Q^{\gamma(n+1)}_{\gamma(n)}$) more or less as in Theorem 3.2. Since each of the indices $\gamma(n)$, $n<\omega$, lies in N, our present approach of treating these indices as variables in the functions of M does not cause any problems. Hence we can finish just as before. □

§4. COLORING LADDER SYSTEMS

Let $S \subseteq \Omega$. A ladder system on S is a sequence $\eta = \langle \eta_\delta \mid \delta \in S \rangle$ such that η_δ is a strictly increasing ω-sequence with limit δ, for each $\delta \in S$. A coloring of η is a sequence $c = \langle c_\delta \mid \delta \in S \rangle$ such that $c_\delta \in {}^{\omega}2$. (We think of $c_\delta(i)$ as coloring $\eta_\delta(i)$.) A uniformisation of c is a function $f: \omega_1 \to 2$ such that for each $\delta \in S$ there is an $n \in \omega$ such that $m \geq n \to c_\delta(m) = f(\eta_\delta(m))$.

If MA + ¬CH is assumed, every coloring of every ladder system is uniformisable. If $\Diamond(S)$ holds, any given ladder system on S will have a non-uniformisable coloring. If $2^{\aleph_0} < 2^{\aleph_1}$ and S contains a club set, any ladder system of S has a non-uniformisable coloring (See [DeSh] for details.)

Shelah has shown (see [Ek])that if there is a stationary set $S \subseteq \Omega$ and a ladder system η on S such that every coloring of η is uniformisable, then there is a non-free Whitehead group of order $\aleph_1$. This makes the consistency of the aforementioned hypothesis with GCH of great interest. In this section we prove the following result:

Theorem 4.1

Assume GCH. Let $S \subseteq \Omega$ be stationary and costationary. Let

$\eta = \langle \eta_\delta \mid \delta \in S \rangle$ be a ladder system on S. There is a poset $\mathbb{P}$ such that:

(i) $|\mathbb{P}| = \aleph_2$;

(ii) $\mathbb{P}$ satisfies the $\aleph_2$-c.c;

(iii) $\mathbb{P}$ is σ-dense;

(iv) $\|GCH\|^{\mathbb{P}} = \mathbf{1}$;

(v) $\|\check{S}$ is stationary $\|^{\mathbb{P}} = \mathbf{1}$

(vi) $\|$Every coloring of η is uniformisable $\|^{\mathbb{P}} = \mathbf{1}$. □

$\mathbb{P}$ will be the limit of a countable support iteration, $\langle \mathbb{P}_\nu | \nu<\omega_2 \rangle$. Since we shall have $|\mathbb{P}_\nu| = \aleph_1$ for all $\nu<\omega_2$, (i) and (ii) will be immediate, and (iv) will then follow by virtue of (iii). Each $\mathbb{P}_\nu$ will be proper, so (v) will be immediate. Hence all we need check is that (iii) and (vi) hold, and that each $\mathbb{P}_\nu$ is proper.

We begin by describing the basic forcing poset. Let $c = \langle c_\delta \mid \delta \in S \rangle$ be a coloring of η. Let $\mathbb{P}(c)$ be the set of all functions f such that dom(f) is an ordinal $\alpha<\omega_1$, ran(f) $\subseteq 2$, and for all $\delta \in S \cap \alpha$ there is an $n<\omega$ such that $m \geq n \rightarrow c_\delta(m) = f(\eta_\delta(m))$. Partially order $\mathbb{P}(c)$ by reverse inclusion. Clearly,

$$\|\text{The coloring } c \text{ of } \eta \text{ is uniformisable}\|^{\mathbb{P}(c)} = \mathbf{1} .$$

Indeed, if G is a V-generic subset of $\mathbb{P}(c)$, then in V[G], $\bigcup G$ is a uniformising function for c on η. Moreover, assuming GCH, $|\mathbb{P}(c)| = \aleph_1$.

<u>Lemma 4.2</u>

$\mathbb{P}(c)$ is proper.

Proof: We utilize Lemma 2.5. Let λ be regular with $\mathbb{P} = \mathbb{P}(c) \in H_\lambda$. Let C be the set of all countable $N \prec H_\lambda$ such that $\mathbb{P} \in N$ and $N = \bigcup_{n<\omega} N_n$, where

$$\mathbb{P} \in N_0 \prec N_1 \prec N_2 \prec \ldots \prec N$$

and $\omega_1 \cap N_n \in N$ for each $n<\omega$. Clearly, C is club in $[H_\lambda]^{\aleph_0}$. We show that $N \prec_{\mathbb{P}} H_\lambda$ for all $N \in C$. So, given $N \in C$ and $p \in \mathbb{P} \cap N$ we must produce a $q \in \mathbb{P}$, $q \leq p$, such that q is $(\mathbb{P},N)$-generic. In fact we do more: we produce a $q \in P$, $q \leq p$, such that the set $G = \{r \in \mathbb{P} \cap N \mid q \leq r\}$ is N-generic on $\mathbb{P} \cap N$ (i.e. if $D \in N$ is dense in $\mathbb{P}$, then $G \cap (D \cap N) \neq \phi$). Let $\alpha = \omega_1 \cap N$. Let N_n, $n<\omega$, be as above, with $\alpha_0 < \alpha_1 < \ldots < \alpha$, where $\alpha_n = \omega_1 \cap N_n$.

<u>Case 1</u>. $\alpha \notin S$.

Let G be any N-generic subset of $\mathbb{P} \cap N$ containing p. Set $q = \cup G$. By N-genericity (and remember that $N \prec H_\lambda$), $q \in {}^{\alpha}2$. Since $\alpha \notin S$, $q \in \mathbb{P}$. So we are done, q being as required.

<u>Case 2</u> $\alpha \in S$.

Let $\langle D_n \mid n < \omega\rangle$ enumerate all dense subsets of $\mathbb{P}$ lying in N. By discarding various of the N_n's if necessary, we may assume (after reindexing) that $D_n \in N_n$ for each $n<\omega$, and that $p \in N_0$. Now, there are only finitely many $i<\omega$ such that $\mathrm{dom}(p) \leq \eta_\alpha(i) < \alpha_0$, say $i_0 < i_1 < \ldots < i_k$. Working in $N_0 \prec H_\lambda$, we can extend p to a condition $\bar{p}_0$ such that $\mathrm{dom}(\bar{p}_0) = \eta_\alpha(i_k)+1$ and $\bar{p}_0(\eta_\alpha^{(i_j)}) = c_\alpha^{(i_j)}$ for $j = 0,\ldots,k$. Then as $D_0 \in N_0$ is dense in $\mathbb{P}$ we can find a condition $p_0 \in \mathbb{P} \cap N_0$, $p_0 \leq \bar{P}_0$, $p_0 \in D_0$. Similarly, working in N_1 now, we can extend p_0 to a condition $\bar{p}_1$ such that $\alpha_0 \subseteq \mathrm{dom}(\bar{p}_1)$ and $\bar{p}_1(\eta_\alpha(i)) = c_\alpha(i)$ for all i such that $\alpha_0 \leq \eta_\alpha(i) < \alpha_1$, and then extend $\bar{p}_1$ to a condition $p_1 \in D_1 \cap N_1$. Continuing thus we define p_n, $n<\omega$, so that $p_n \in D_n \cap N_n$, $p_0 \geq p_1 \geq p_2 \geq \ldots$, $\alpha_n \subseteq \mathrm{dom}(p_{n+1})$, and $p_{n+1}(\eta_\alpha(i)) = c_\alpha(i)$ for all i such that $\alpha_n \leq \eta_\alpha(i) < \alpha_{n+1}$. Let $G = \{q \in \mathbb{P}\cap N \mid (\exists n)(p_n \leq q)\}$. Then G is N-generic on $\mathbb{P}\cap N$ and $q = \bigcup G \in {}^{\alpha}2$. Put $q(\eta_\alpha(i)) = c_\alpha(i)$ for all $i \geq i_0$. Hence $q \in \mathbb{P}$, and we are done. □

An immediate consequence of the above proof is the following result:

Lemma 4.3

$\mathbb{P}(c)$ is σ-dense. □

The iteration sequence $\langle \mathbb{P}_\nu \mid \nu < \omega_2 \rangle$ is a countable support iteration of the posets $\mathbb{P}(c)$, with the colorings c chosen in the usual fashion so as to ensure (vi) of Theorem 4.1. So, in particular, for each $\nu < \omega_2$ we shall have an element $c^\nu \in V^{(\mathbb{P}_\nu)}$ such that $\|c^\nu$ is a coloring of $\check{\eta}\|^{\mathbb{P}_\nu} = \mathbb{1}$ and $\mathbb{P}_{\nu+1} = \mathbb{P}_\nu \otimes \mathbb{P}(c^\nu)$. It will then be an immediate consequence of Lemma 4.2 and Theorem 3.3 that each $\mathbb{P}_\nu$ is proper, $\nu < \omega_2$. But wait a minute, will it not be the case that because of all of the boolean names which occur in a forcing product $\mathbb{P}_\nu \otimes \mathbb{P}(c^\nu)$ that eventually (or indeed very quickly) $\|\mathbb{P}_\nu\| > \aleph_1$? Well, it all depends upon the precise definition of the product poset. According to the definition adopted in §0, already $\mathbb{P}_1$ could be too large. However, we shall not run into this problem, since for $\mathbb{P}_{\nu+1}$ we shall not take all of $\mathbb{P}_\nu \otimes \mathbb{P}(c^\nu)$, but rather a certain dense subset of this product. This ties in with our proof of part (iii) of Theorem 4.1, the only part left to prove now. It suffices to show that $\mathbb{P}_\nu$ is σ-dense for each $\nu < \omega_2$.

Let $\langle \mathbb{P}_\nu | \nu < \omega_2 \rangle$ be the iteration sequence defined in the standard fashion, with $\mathbb{P}_{\nu+1} \cong \mathbb{P}_\nu \otimes \mathbb{P}(c^\nu)$ for each $\nu < \omega_2$. Let $\bar{\mathbb{P}}_\nu$, $\nu < \omega_2$, consist of all functions p such that $\mathrm{dom}(p) \subseteq \nu$, $|\mathrm{dom}(p)| \leq \aleph_0$, and for all $\zeta \in \mathrm{dom}(p)$, $p(\zeta)$ is a function such that $p \restriction \zeta \in \bar{\mathbb{P}}_\zeta$ and

$$p \restriction \zeta \Vdash_{\bar{\mathbb{P}}_\zeta} \text{"}\overset{\smile}{p(\zeta)} \in \mathbb{P}(c^\zeta)\text{"}.$$

The ordering of $\bar{\mathbb{P}}_\nu$ is

$$p \leq q \text{ iff } \mathrm{dom}(p) \supseteq \mathrm{dom}(q) \ \& \ (\forall \zeta \in \mathrm{dom}(q))(p(\zeta) \supseteq q(\zeta)).$$

Clearly, $\bar{\mathbb{P}}_\nu \subseteq \mathbb{P}_\nu$ for all $\nu < \omega_2$. Moreover, $|\mathbb{P}_\nu| = \aleph_1$ for all $\nu < \omega_2$. We shall finish our proof by showing that $\bar{\mathbb{P}}_\nu$ is dense in $\mathbb{P}_\nu$ for all $\nu < \omega_2$ and that each $\bar{\mathbb{P}}_\nu$ is σ-dense.

<u>Lemma 4.4</u>

(A) $\bar{\mathbb{P}}_\nu$ is dense in $\mathbb{P}_\nu$, and indeed if $p \in \mathbb{P}_\nu$ there is a $p^* \in \mathbb{P}$, $p^* \leq p$, such that for some $\delta \in \Omega - S$, $(\forall \zeta \in \mathrm{dom}(p^*))(\mathrm{dom}(p^*(\zeta)) = \delta)$.

(B) $\bar{\mathbb{P}}_\nu$ is σ-dense.

Proof: By a simultaneous induction on ν. Once we have (A) for ν, we can work with conditions of the type p^* in order to repeat the proof of lemma 4.2, Case 1, in order to show that $\bar{\mathbb{P}}_\nu$ is σ-dense. So what we need to prove is that if (A) and (B) hold <u>below ν,</u> then (A) holds <u>at ν</u>. Well, it is easily seen that if (A) and (B) hold below ν, then whenever $p \in \mathbb{P}_\nu$ and $\zeta \in \mathrm{dom}(p)$, there is a $q \in \bar{\mathbb{P}}_\zeta$, $q \leq p\restriction\zeta$, and a function $q(\zeta)$, such that $q \Vdash_{\mathbb{P}_\zeta}$ "$\check{q(\zeta)} = p(\zeta)$".

Let $N = \langle H_\lambda, \in, \mathbb{P}_\nu, \Vdash_{\mathbb{P}_\nu}\rangle$, where λ is large and regular, and let $\langle N_\delta | \delta<\omega\rangle$ be a strictly increasing sequence of countable elementary submodels of N, continuous at limits, with $p \in N_0$ and $\delta \subseteq N_\delta$. Let $C \subseteq \omega_1$ be club in ω_1 with $N_\delta \cap \omega_1 = \delta$ for all $\delta \in C$, and let δ be a limit point of C which is not in S. (Since S is costationary in ω_1, this is in order.) Let $\langle \delta_n | n<\omega\rangle$ be a strictly increasing sequence of members of C cofinal in δ. Now, $N_{\delta_n} \prec H_\lambda$ for each $n < \omega$, so we may make repeated use of the extension property mentioned above in order to pick conditions $p_n \in N_{\delta_n} \cap \mathbb{P}_\nu$, such that $p \geq p_0 \geq p_1 \geq \ldots$, and ordinals $\beta_n \in \text{dom}(p_n)$, so that $p_n\restriction\beta_n \in \bar{\mathbb{P}}_{\beta_n}$, $p_n\restriction\beta_n \Vdash$ "$p(\beta_n) = \check{x}$" for some function x, $\delta_n \subseteq \text{dom}(p_{n+1}(\beta_n))$, and such that each element of $\bigcup_{n<\omega} \text{dom}(p_n)$ will be β_m for infinitely many $m < \omega$. It is easily seen that $p^* = \bigcup_{n<\omega} p_n$ is as required now. □

§5. THE PROPER FORCING AXIOM

Let PFA (Proper Forcing Axiom) be the following statement: if $\mathbb{P}$ is a proper poset and $\mathfrak{J}$ is a collection of $\aleph_1$ many dense subsets of $\mathbb{P}$, then $\mathbb{P}$ has an $\mathfrak{J}$-generic subset.

Since every c.c.c. poset is proper, PFA is a generalisation of MA_{ω_1}. As we shall see, it is much stronger than MA_{ω_1}.

We commence by establishing the consistency of PFA with ZFC + $2^{\aleph_0} = \aleph_2$. The obvious idea is to commence with a model of ZFC + GCH, and iterate proper poset forcing in a manner analogous to the usual consistency proof for MA_{ω_1}. By Theorems 3.1 and 3.3, the iteration will be proper. So, in particular, stationary subsets of ω_1 will remain stationary, which means that $\aleph_1$ will be preserved. (Since $\aleph_1$ is the only parameter involved in PFA, it will not matter if other cardinals are collapsed in the iteration, so long as we end up with a model of PFA.) The problem with PFA, however, is how long the iteration needs to be. For MA_{ω_1}, if we start out with GCH, then an iteration of length ω_2 suffices. This is because, although MA_{ω_1} refers to all c.c.c. posets (of which there is a proper class, of course), by a simple Löwenheim-Skolem argument it can be shown that the full MA_{ω_1} is a consequence of MA_{ω_1} restricted to posets of cardinality at most $\aleph_1$. Hence in our iteration, we need only deal with these posets, of which there are only $2^{\aleph_0} = \aleph_2$, even allowing for the new posets which arise during the course of the iteration. But with PFA we have no such reduction. Certainly, if we only require PFA for proper posets of cardinality at most $\aleph_1$, then we can achieve this by a straightforward iteration entirely parallel to that for MA_{ω_1}, only using a countable support iteration, of course. [There is a little difficulty encountered in proving that the iteration satisfies the $\aleph_2$-c.c. Even if we start by assuming GCH, which we do, this is not quite immediate, for if $\langle \mathbb{P}_\nu \mid \nu \leq \omega_2 \rangle$ is the countable support iteration obtained by iterating all proper posets of cardinality $\aleph_1$, then it is not the case that $|\mathbb{P}_\nu| = \aleph_1$ for all $\nu < \omega_2$, because of all the boolean names which we allow in product posets. However, it is easily seen that if $\mathbb{P}$ is a proper poset, any countable set of ordinals (say) in $V^{(\mathbb{P})}$ is contained in a countable set of ordinals of V, and using this fact it is easy to prove, by induction, that for every $\nu < \omega_2$, $\mathbb{P}_\nu$ has a dense subset $\mathbb{R}_\nu$ of cardinality at most $\aleph_1$. Now the proof that $\mathbb{P}_{\omega_2}$ has the $\aleph_2$-c.c. is straightforward.] But for a full PFA we require some device to restrict our attention from the proper class of all proper posets to some representative set of such posets. This can be done, but at the cost of a large cardinal assumption (and a significant one at that). Moreover, it can be shown that some such assumption is necessary, since PFA implies the existence of inner models of set theory with various large cardinals (measurable, and beyond).

Recall that a cardinal κ is <u>supercompact</u> iff for every $\lambda \geq \kappa$ there is an ultrafilter U on $[\lambda]^{<\kappa}$ such that:

(i) the ultrapower $V^{[\lambda]^{<\kappa}}/U$ is well-founded;

and, if M is the transitive collapse of $V^{[\lambda]^{<\kappa}}/U$, then

(ii) $[M]^\lambda \subseteq M$;

(iii) if $j : V \prec M$ is the canonical embedding, then

$$j \upharpoonright V_\kappa = id \upharpoonright V_\kappa \text{ and } j(\kappa) > \lambda.$$

(We refer to such a U as a <u>super-compact ultrafilter</u> on $[\lambda]^{<\kappa}$, and the embedding j as a <u>(κ,λ)-embedding</u>.)

The following result of R. Laver is the key device we need in order to get a model of PFA:

Theorem 5.1

Let κ be a supercompact cardinal. Then there is a function $f : \kappa \to V_\kappa$ such that for any set x, if $\lambda \geq |TC(x)|$, then there is a supercompact ultrafilter on $[\lambda]^{<\kappa}$ such that if j is the associated (κ,λ)-embedding, then $[j(f)](\kappa) = x$.

Proof: Suppose not. Then for each $f : \kappa \to V_\kappa$ there is a least ordinal λ_f such that there is an x with $|TC(x)| \leq \lambda_f$ and $\langle x, \lambda_f \rangle$ a counterexample for f. Let ν exceed all the λ_f's and pick a supercompact ultrafilter U_ν on $[\nu]^{<\kappa}$.

Let $\Phi(g,\delta)$ be the statement that for some cardinal α, $g:\alpha \to V_\alpha$, and δ is the least cardinal for which there is an x with $|TC(x)| \leq \delta$, such that for no supercompact ultrafilter U_δ on $[\delta]^{<\alpha}$ does $[j_\delta(g)](\alpha) = x$.

Notice that as $[M_\nu]^\nu \subseteq M_\nu$, $M_\nu \models \Phi(f,\lambda_f)$ for all $f : \kappa \to V_\kappa$, where M_ν is the collapse of $V^{[\nu]^{<\kappa}}/U_\nu$.

Let U_κ be the projection of U_ν onto κ. Then U_κ is a measure ultrafilter on κ and there is a canonical embedding $h_{\kappa,\lambda} : V^\kappa/U_\kappa \prec V^{[\lambda]^{<\kappa}}/U_\nu$. Consequently, we see that there is a set $A \in U_\kappa$ such that for each $\alpha \in A$ and each $f' : \alpha \to V_\alpha$ there is a $\lambda_{f'} < \kappa$ such that $\Phi(f',\lambda_{f'})$. (Because the same is true of κ.)

Define $f : \kappa \to V_\kappa$ by recursion, as follows. Suppose $\alpha < \kappa$ and $f_\alpha = f \upharpoonright \alpha$ is defined. Let $f(\alpha) = \phi$ unless $\alpha \in A$ and $f_\alpha : \alpha \to V_\alpha$. In this case there is an $x \in V_\kappa$ witnessing $\Phi(f_\alpha, \lambda_{f_\alpha})$. Let x_α be such an x, and define $f(\alpha) = x_\alpha$.

Notice that, if j_ν is the (κ,ν)-embedding associated with U_ν, then:

$$[j_\nu(\langle f_\alpha | \alpha \in A\rangle)](\kappa) = f;$$

$$[j_\nu(\langle \lambda_{f_\alpha} | \alpha \in A\rangle)](\kappa) = \lambda_f;$$

$$[j_\nu(\langle x_\alpha \mid \alpha \in A\rangle)](\kappa) = [j_\nu(f)](\kappa) = x,$$

where x is a witness to $\Phi(f,\lambda_f)$ in M_ν, and hence in V.

Let U_{λ_f} be the projection of U_ν onto λ_f. The following diagram commutes and is elementary (where $h_{\lambda_f\nu}$ is the canonical embedding of M_{λ_f} into M_ν):

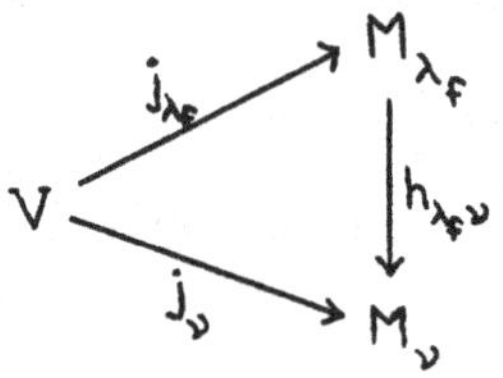

We have $h_{\lambda_f\nu} = \mathrm{id}\restriction\lambda_f$, $|TC(x)| \leq \lambda_f$; so $x \in M_{\lambda_f}$ and $h_{\lambda_f\nu}(x) = x$. Thus:

$$[j_{\lambda_f}(f)](\kappa) = [(h_{\lambda_f\nu})^{-1}]((j_\nu(f))(\kappa)) = (h_{\lambda_f\nu})^{-1}(x) = x.$$

But this contradicts the fact that x witnesses $\Phi(f,\lambda_f)$, so we are done. □

We are now ready to prove our consistency result for PFA.

<u>Theorem 5.2</u>

Con(ZFC+ "there is a supercompact cardinal")

$\rightarrow$ Con(ZFC+ $2^{\aleph_0} = \aleph_2$ + PFA).

Proof: Let κ be a supercompact cardinal, and let $f: \kappa \rightarrow V_\kappa$ be as in Theorem 5.1. We call a pair $(\mathbb{P},\mathfrak{Z})$ <u>suitable</u> if $\mathbb{P}$ is a proper poset and $\mathfrak{F}$ is a sequence of length less than κ of dense subsets of $\mathbb{P}$.

We shall construct a countable support iteration sequence $\langle \mathbb{P}_\alpha \mid \alpha \leq \kappa \rangle$ of proper posets as follows. Let $\mathbb{P}_0$ be the usual poset for adding a Cohen generic subset of ω. If $\mathbb{P}_\alpha$ is defined and $f(\alpha) \in V^{(\mathbb{P}_\alpha)}$ is such that $\Vdash_{\mathbb{P}_\alpha}$ " $f(\alpha) = (\mathbb{P}, \mathfrak{F})$ is a suitable pair", then let $\mathbb{P}_{\alpha+1} = \mathbb{P}_\alpha \otimes \mathbb{P}$. If $\mathbb{P}_\alpha$ is defined and $f(\alpha)$ is not of the above form, let $\mathbb{P}_{\alpha+1} = \mathbb{P}_\alpha \otimes \mathbb{P}_0$. We shall show that $\Vdash_{\mathbb{P}_\kappa}$ "$2^{\aleph_0} = \aleph_2 = \kappa$ & PFA".

A routine Δ-system argument shows that $\mathbb{P}_\kappa$ has the κ-c.c. Hence κ remains a cardinal after $\mathbb{P}_\kappa$-forcing. Moreover, since $|\mathbb{P}_\kappa| = \kappa$, we have $\Vdash_{\mathbb{P}_\kappa}$ "$2^{\aleph_0} \leq \kappa$". Since $\mathbb{P}_\kappa$ is proper, ω_1 remains a cardinal under $\mathbb{P}_\kappa$-forcing. So, as PFA $\rightarrow$ MA_{ω_1} $\rightarrow$ $2^{\aleph_0} > \omega_1$, it remains only to show that $\Vdash_{\mathbb{P}_\kappa}$ "$\kappa \leq \omega_2$ & PFA". It suffices to show that if G is V-generic on $\mathbb{P}_\kappa$ and $(\mathbb{P}, \mathfrak{F})$ is a suitable pair in V[G], then there is a filter $F \subseteq \mathbb{P}$ in V[G] such that $F \cap D \neq \emptyset$ for every $D \in \mathfrak{F}$. For, this clearly means that PFA holds in V[G]. And moreover, on the assumption that $\kappa > \omega_2$ in V[G], it implies that there is a function from ω_1 onto ω_2 in V[G], which is absurd. (Consider the proper poset of all countable maps from ω_1 to ω_2, together with the appropriate collection of dense sets.)

So let G be a V-generic subset of $\mathbb{P}_\kappa$, and let $(\mathbb{P}, \mathfrak{F}) \in V[G]$ be (in V[G]) a suitable pair. We may assume that the domain of $\mathbb{P}$ is some cardinal λ. Let $\mathring{\mathbb{P}}, \mathring{\mathfrak{F}}$ be $\mathbb{P}_\kappa$-names for $\mathbb{P}, \mathfrak{F}$, and suppose that $\Vdash_{\mathbb{P}_\kappa}$ "$(\mathring{\mathbb{P}},\mathring{\mathfrak{F}})$ is a suitable pair". (As usual this causes no loss of generality.) By the choice of κ, f, we can find a (κ,λ)-embedding $j : V \prec M$ such that $[j(f)](\kappa) = (\mathring{\mathbb{P}},\mathring{\mathfrak{F}})$.

Now, by construction, $\mathbb{P}_\kappa \subseteq V_\kappa$. Hence $j \restriction \mathbb{P}_\kappa = \mathrm{id} \restriction \mathbb{P}_\kappa$ and $\mathbb{P}_\kappa \subseteq j(\mathbb{P}_\kappa)$. Moreover, by elementarity, $j(\mathbb{P}_\kappa)$ is constructed from $j(f)$ in M exactly as $\mathbb{P}_\kappa$ is constructed from f in V. But $[j(f)](\alpha) = f(\alpha)$ for all $\alpha < \kappa$. Hence $\mathbb{P}_\kappa = [j(\mathbb{P}_\kappa)]_\kappa = \{p \restriction \kappa \mid p \in j(\mathbb{P}_\kappa)\}$, and if we set $j(\mathbb{P}_\kappa)_\kappa^{j(\kappa)} = \{p \restriction (j(\kappa)-\kappa) \mid p \in j(\mathbb{P}_\kappa)\}$, then $j(\mathbb{P}_\kappa) \cong \mathbb{P}_\kappa \otimes j(\mathbb{P}_\kappa)_\kappa^{j(\kappa)}$.

Since $M \subseteq V$, G is M-generic on $\mathbb{P}_\kappa$. So if we choose H to be M[G]-generic on $j(\mathbb{P}_\kappa)_\kappa^{j(\kappa)}$, then $G \otimes H$ will be M-generic on $j(\mathbb{P}_\kappa)$.

<u>Claim</u> : The embedding $j : V \prec M$ extends to an embedding $j^* : V[G] \prec M[G \otimes H]$.

Proof of claim : Let $x \in V[G]$. Then for some term τ and some $a_0, \ldots, a_n \in V$, $x = \tau^{V[G]}(a_0, \ldots, a_n, G)$. Let $j^*(x) = \tau^{M[G \otimes H]}(j(a_0), \ldots, j(a_n), G \otimes H)$. That j^* is well-defined will follow from our proof that j^* is elementary.

Suppose $V[G] \models \Phi(\tau_0^{V[G]}(\vec{x}_0, G), \ldots, \tau_n^{V[G]}(\vec{x}_n, G))$, where $\vec{x}_i \in V$. Then there is a $p \in G$ such that $p \Vdash^V_{\mathbb{P}_\kappa} \Phi(\tau_0(\mathring{\vec{x}}_0, \mathring{G}), \ldots, \tau_n(\mathring{\vec{x}}_n, \mathring{G}))$.

So applying j, $j(p) \Vdash^M_{j(\mathbb{P}_\kappa)} \Phi(\tau_0(j(\mathring{\vec{x}}_0), \mathring{G}), \ldots, \tau_n(j(\mathring{\vec{x}}_n), \mathring{G}))$. But $p \in \mathbb{P}_\kappa$, so $j(p) = p \in G \subseteq G \otimes H$. Hence $M[G \otimes H] \models \Phi(\tau_0^{M[G \otimes H]}(j(\vec{x}_0), G \otimes H), \ldots, \tau_n^{M[G \otimes H]}(j(\vec{x}_n), G \otimes H))$, i.e.

$M[G \otimes H] \models \Phi(j^*(\tau_0^{V[G]}(\vec{x}_0,G)),\ldots j^*(\tau_n^{V[G]}(\vec{x}_n,G)))$. Since we can carry out exactly the same argument with $\neg\Phi$ in place of Φ, the claim is proved. Δ

Now, $j\restriction\lambda \subseteq M$ and $|j\restriction\lambda| = \lambda$. Hence as $[M]^\lambda \subseteq M$, we have $j\restriction\lambda \in M$. Thus $j\restriction\lambda \in M[G \otimes H]$. Consider the action of $j\restriction\lambda$ on $(\mathbb{P},\mathfrak{F})$. Clearly, $j\restriction\lambda$ embeds $\mathbb{P}$ in $j^*(\mathbb{P})$ (in an order preserving fashion), and for each $\nu<\text{otp}(\mathfrak{F})$, $j\restriction\lambda$ embeds $\mathfrak{F}_\nu$ into the set $j^*(\mathfrak{F})_\nu$. (This is an immediate consequence of $j^*: V[G] \prec M[G \otimes H]$ and $j^*\restriction\lambda = j\restriction\lambda$.) Thus, using an obvious terminology, which we shall assume implies the suitability of both pairs, $M[G \otimes H] \models$ "$j\restriction\lambda$ embeds $(\mathbb{P},\mathfrak{F})$ in $(j^*(\mathbb{P}),j^*(\mathfrak{F}))$".

Now let K be any M-generic subset of $j(\mathbb{P}_\kappa)$. Then $K = G_K \otimes H_K$, where G_K is M-generic on $\mathbb{P}_\kappa$ and H_K is $M[G_K]$-generic on $j(\mathbb{P}_\kappa)_\kappa^{j(\kappa)}$. Since $j\restriction V_\kappa = id\restriction V_\kappa$ and $[M]^\lambda \subseteq M$, it follows that G_K is in fact V-generic on $\mathbb{P}_\kappa$. Hence the above argument is still valid, and so

$$M[K] \models \text{"}j\restriction\lambda \text{ embeds } (\mathbb{P},\mathfrak{F}) \text{ in } (j^*(\mathbb{P}),j^*(\mathfrak{F}))\text{"}.$$

Thus as K is arbitrary as above,

$$\Vdash^M_{j(\mathbb{P}_\kappa)} \text{"} \widetilde{j\restriction\lambda} \text{ embeds } (\mathring{\mathbb{P}},\mathring{\mathfrak{F}}) \text{ in } (j(\mathring{\mathbb{P}}), j(\mathring{\mathfrak{F}}))\text{"}.$$

But $(\mathring{\mathbb{P}},\mathring{\mathfrak{F}}) = [j(f)](\kappa)$. Hence

(i) $M \models [j(f)](\kappa)$ is a pair of members of $V^{j(\mathbb{P}_\kappa)_\kappa}$;

(ii) $\Vdash^M_{j(\mathbb{P})_\kappa}$ "$[j(f)](\kappa)$ is embeddable in $(j(\mathring{\mathbb{P}}),j(\mathring{\mathfrak{F}}))$".

Hence:

(iii) $M \models (\exists\alpha<j(\kappa))[[j(f)](\alpha)$ is a pair of members of $V^{j(\mathbb{P}_\kappa)_\alpha}$ and $\Vdash_{j(\mathbb{P}_\kappa)}$"$[j(f)](\alpha)$ is embeddable in $(j(\mathring{\mathbb{P}}),j(\mathring{\mathfrak{F}}))$"$]$

Applying $j : V \prec M$ gives:

(iv) $V \models (\exists\alpha<\kappa)$ [$f(\alpha)$ is a pair of members of $V^{(\mathbb{P}_\alpha)}$ and

$\Vdash_{\mathbb{P}_\kappa}$ " $f(\alpha)$ is embeddable in $(\mathring{\mathbb{P}},\mathring{\mathfrak{F}})$"].

(where we have used the fact that $(\mathbb{P}_\kappa)_\alpha = \mathbb{P}_\alpha$).

Pick some $\alpha<\kappa$ witnessing (iv). Now, since we have adopted the convention that embeddability implies the two pairs are suitable, by (iv) we have

$$\Vdash_{\mathbb{P}_\kappa} \text{"}f(\alpha) \text{ is a suitable pair"}.$$

But $f(\alpha)$ is a pair of members of $V^{(\mathbb{P}_\alpha)}$, so clearly,

$$\Vdash_{\mathbb{P}_\alpha} \text{"}f(\alpha) \text{ is a suitable pair"}.$$

Hence by construction, $G_\alpha^{\alpha+1} = G \upharpoonright \mathbb{P}_\alpha^{\alpha+1}$ is $V[G_\alpha]$-generic on $(f(\alpha))_0$. In $V[G]$ now, let $\pi : (f(\alpha)) \to (\mathbb{P},\mathfrak{F})$ be an embedding. (Such a π exists by virtue of (iv).) Then, clearly, $\pi[G_\alpha^{\alpha+1}]$ is an $\mathfrak{F}$-generic filter on $\mathbb{P}$, and we are done. [Notice in particular that since $G_\alpha^{\alpha+1}$ meets every dense subset of $(f(\alpha))_0$ in $V[G_\alpha]$, it certainly meets every member of $(f(\alpha))_1$, and hence $\pi[G_\alpha^{\alpha+1}]$ meets every member of $\mathfrak{F}$.] □

We now give a few sample applications of PFA. We commence with two applications in tree theory, both due to Baumgartner.

Let T be a tree of height ω_1 and cardinality $\aleph_1$. We say T is <u>special</u> if there is a function $f : T \to \omega$ such that whenever $s,t,u \in T$, if $f(s) = f(t) = f(u)$ and $s \leq t,u$, then $t \lesssim u$. In case T has no ω_1-branches, this is equivalent to saying that there is an $h : T \to \omega$ such that $h(s) = h(t)$ implies that s and t are incomparable in T. [One implication is totally trivial. For the other, given f as above, each set $f^{-1}[n]$ necessarily consists of totally incomparable chains, each of which will be countable, so we can partition $f^{-1}[n]$ into a countable family of antichains, and thereby obtain h.]

Lemma 5.3

Let T be a tree of height ω_1 and cardinality $\aleph_1$. If T is special, then T has at most $\aleph_1$ ω_1-branches.

Proof: Let $f : T \to \omega$ specialise T. If B is an ω_1-branch of T, we can find an $n \in \omega$ such that $\{x \in B \mid f(x) = n\}$ is uncountable. Choose any $t \in B$ such that $f(t) = n$. Then, clearly,

$$B = \{x \in T \mid (x \leq t) \text{ or } (x > t \text{ and } (\exists y > x)(f(y) = n))\}.$$

Hence B is determined by t and n. The lemma follows. □

Lemma 5.4

Let T be a tree of height ω_1 and cardinality $\aleph_1$. If T has at most $\aleph_1$ ω_1-branches, then there is a c.c.c. poset Q such that $\Vdash_Q$ "T is special".

Proof: By enlarging T if necessary, we may assume that T has exactly $\aleph_1$ ω_1-branches, and that every element of T belongs to an ω_1-branch. Let $\langle B_\alpha \mid \alpha < \omega_1 \rangle$ enumerate the ω_1-branches of T. For each $\alpha < \omega_1$, let $s(\alpha)$ be the best element of B_α not in $\bigcup_{\beta<\alpha} B_\beta$, and set $S = \{s(\alpha) \mid \alpha < \omega_1\}$. Thus $|B_\alpha \cap S| \leq \aleph_0$ for all $\alpha < \omega_1$. In particular, there is no ω_1-branch through S, where we regard S as a subtree of T.

Suppose that $f : S \to \omega$ is such that if $f(s) = f(t)$, then s and t are incomparable in S. Then we can define $g : T \to \omega$ by the following. Let $t \in T$. Let α be least such that $t \in B_\alpha$. Thus $t \geq s(\alpha)$. Let $g(t) = f(s(\alpha))$. It is easily checked that g specialises T. Hence it suffices to find a c.c.c. poset Q such that

$\Vdash_Q$" $\mathring{f} : S \to \omega$ is such that if $\mathring{f}(S) = \mathring{f}(t)$, then s and t are S-incomparable".

Let Q consist of all finite functions p from S to ω such that if $p(s) = p(t)$, then s and t are S-incomparable. Let $p \leq q$ iff $p \supseteq q$. Providing it satisfies the c.c.c., Q will clearly be as required. So assume not, and let $J = \{p_\alpha | \alpha<\omega_1\}$ be an uncountable, pairwise incompatible set. W.l.o.g. we may assume that $|p_\alpha| = n$ for all $\alpha<\omega_1$, where n is minimal such. We may assume further that if $\alpha \neq \beta$, then $\mathrm{dom}(p_\alpha) \cap \mathrm{dom}(p_\beta) = \phi$, and that if $\alpha<\beta$, $p_\alpha(s) = p_\beta(t)$, $s \neq t$, and $s \lessgtr t$, then in fact $s < t$. (Notice that since p_α and p_β are incompatible, there always are such s,t.) Let U be a uniform ultrafilter on ω_1. For each $\alpha \in \omega_1$, let $\mathrm{dom}(p_\alpha) = \{s_0^\alpha, \ldots, s_{n-1}^\alpha\}$. For each α there will be $i(\alpha), j(\alpha)<n$ such that

$$\{\beta \in \omega_1 \mid s^\alpha_{i(\alpha)} < s^\beta_{j(\alpha)}\} \in U .$$

Furthermore, there must be $i, j < n$ such that

$$A = \{\alpha \mid i(\alpha) = i \ \& \ j(\alpha) = j\} \in U.$$

Let $\alpha_1, \alpha_2 \in A$. Then for some $\beta > \alpha_1, \alpha_2$, $s_i^\alpha < t_j^\beta$ for $\alpha = \alpha_1, \alpha_2$. Thus $s_i^{\alpha_1}$ and $s_i^{\alpha_2}$ are comparable. Thus $\{s_i^\alpha \mid \alpha \in A\}$ determines an ω_1-branch through S, a contradiction. The proof is complete.□

Lemma 5.5

Assume $2^{\aleph_0} \geq \aleph_2$. Let T be a tree of height ω_1 and cardinality $\aleph_1$. Let $\mathbb{P}$ be a countably closed poset. Then $\Vdash_{\mathbb{P}}$ "All ω_1-branches of $\check{T}$ lie in $\check{V}$".

Proof: Suppose, on the contrary, that

$$\Vdash_{\mathbb{P}} \text{"B is an } \omega_1\text{-branch of } \check{T} \text{ and } B \notin \check{V}\text{"}.$$

Since $\Vdash_{\mathbb{P}}$ "$B \notin \check{V}$", it is easy to construct, inductively, elements p_f of $\mathbb{P}$, x_f of T, for $f \in 2^{\underset{\smile}{\omega}}$, such that:

(i) $f \subseteq g \rightarrow p_g \leq p_f$ & $x_f \leq x_g$;

(ii) $p_f \Vdash_{\mathbb{P}}$ "$\check{x}_f \in B$";

(iii) $x_{f^\frown\langle 0\rangle} \nparallel x_{f^\frown\langle 1\rangle}$.

Since $\mathbb{P}$ is countably closed, for each $f \in 2^{\omega}$ we can find a $p_f \in \mathbb{P}$ such that $p_f \leq p_{f\restriction n}$ for all $n < \omega$. Then $p_f \Vdash_{\mathbb{P}}$ "$\{x_{f\restriction n} | n<\omega\} \subseteq B$; so for some $p_f' \leq p_f$ and some $y_f \in T$, we have $p_f' \Vdash_{\mathbb{P}}$ "$\check{y}_f \in B$ & $(\forall n<\omega)(\check{y}_f > \check{x}_{f\restriction n})$". Then $f \neq g$ implies $y_f \nparallel y_g$, so $|T| \geq 2^{\aleph_0} > \aleph_1$. Contradiction. □

Theorem 5.6

Assume PFA. If T is a tree of height ω_1 and cardinality $\aleph_1$, then T is special. Hence every tree of height ω_1 and cardinality $\aleph_1$ has at most $\aleph_1$ ω_1-branches.

Proof. Suppose T has λ ω_1-branches, where $\lambda \geq \aleph_1$. Let $\mathbb{P}$ be the poset of all countable maps from ω_1 to λ, ordered by reverse inclusion. Then $\mathbb{P}$ is proper and $\Vdash_{\mathbb{P}}$ "$|\lambda| = \aleph_1 = \check{\aleph}_1$". So by lemma 5.5 (which is valid because PFA $\rightarrow$ $MA_{\omega_1} \rightarrow 2^{\aleph_0} \geq \aleph_2$), $\Vdash_{\mathbb{P}}$ "$\check{T}$ has $\aleph_1$ ω_1-branches". In $V^{(\mathbb{P})}$, choose Q a c.c.c. poset so that

$\Vdash_{Q}$ "$\check{T}$ is special", as in lemma 5.4. Now, $\|\mathring{Q}$ is a proper poset $\|^{\mathbb{P}} = \mathbf{1}$. Hence $\mathbb{R} = \mathbb{P} \otimes Q$ is a proper poset (in V). Moreover, for some $\mathring{f} \in V^{(\mathbb{R})}$,

$$\Vdash_{\mathbb{R}} \text{"}\mathring{f} : \check{T} \to \omega \text{ specialises } \check{T}\text{"}.$$

For each $t \in T$, let

$$D_t = \{r \in \mathbb{R} \mid r \Vdash_{\mathbb{R}} \text{"}\mathring{f}(\check{t}) = \check{n}\text{", for some } n \in \omega\}.$$

Clearly, D_t is dense in $\mathbb{R}$. Let

$$\mathcal{F} = \{D_t \mid t \in T\}.$$

By PFA, let G be an $\mathcal{F}$-generic subset of $\mathbb{R}$. Then we can define a function $F : T \to \omega$ by

$$F(t) = n \quad \text{iff} \quad (\exists r \in G)(r \Vdash_{\mathbb{R}} \text{"}\mathring{f}(\check{t}) = \check{n}\text{"}).$$

Clearly F specialises T. □

Theorem 5.7

Assume PFA. Then there are no ω_2-Aronszajn trees.

Proof : Suppose, on the contrary, that T were an ω_2-Aronszajn tree. Let $\mathbb{P}$ be the poset of all countable maps from ω_1 into ω_2, ordered by reverse inclusion. Thus $\Vdash_{\mathbb{P}}$ "$|\check{\omega}_2| = \aleph_1 = \check{\aleph}_1$". Since the levels of T have cardinality at most $\aleph_1$, the same proof as in lemma 5.5 (more or less) shows that $\Vdash_{\mathbb{P}}$ "$\check{T}$ has no branches of type $\check{\omega}_2$". In $V^{(\mathbb{P})}$, let h be a strictly increasing, cofinal map of $\check{\omega}_1$ into $\check{\omega}_2$, and let $S = \bigcup_{\alpha<\omega_1} \check{T}_{h(\alpha)}$. Then, in $V^{(\mathbb{P})}$, S is a tree of height ω_1, cardinality $\aleph_1$, with no ω_1-branches. Let Q be the poset defined in lemma 5.4. Thus there is an $f \in (V^{(\mathbb{P})})^{(Q)}$ such that, with boolean value $\mathbf{1}$, f maps S into ω in such a way that if $f(s) = f(t)$ then s and t are incomparable in S (hence T).

Let $\mathbb{R} = \mathbb{P} \otimes Q$. Since $\mathbb{P}$ is countably closed and Q has the c.c.c. in $V^{(\mathbb{P})}$, $\mathbb{R}$ is proper. We shall apply PFA to $\mathbb{R}$.

For each $\gamma < \omega_2$ (working in V from now on), let $\ell_\gamma = \omega_1 \overset{\text{onto}}{\to} T_\gamma$. For $\alpha, i < \omega_1$, let

$$D_{\alpha i} = \{p \in \mathbb{R} \mid \exists\gamma\exists n [p \Vdash_{\mathbb{R}} "\mathring{h}(\check{\alpha}) = \check{\gamma} \;\&\; \mathring{f}(\widecheck{\ell_\gamma(i)}) = \check{n}"]\}.$$

Set

$$\mathfrak{F} = \{D_{\alpha i} \mid \alpha, i < \omega_1\}.$$

Clearly, $\mathfrak{F}$ is a set of dense subsets of $\mathbb{R}$. Let G be $\mathfrak{F}$-generic on $\mathbb{R}$. Define a function $H : \omega_1 \to \omega_2$ by

$$H(\alpha) = \gamma \quad \text{iff} \quad (\exists p \in G)(p \Vdash_{\mathbb{R}} "\mathring{h}(\check{\alpha}) = \check{\gamma}").$$

Clearly, H is strictly increasing (though not, of course, cofinal!). If we now set $U = \bigcup_{\alpha<\omega_1} T_{H(\alpha)}$, we can define $F : U \to \omega$ by

$$F(u) = n \quad \text{iff} \quad (\exists p \in G)(p \Vdash_{\mathbb{R}} "\mathring{f}(\check{u}) = \check{n}").$$

Clearly, if $F(u) = F(u')$, the u and u' are incomparable in T. Now choose $x \in T_\zeta$, where $\xi = \sup(\operatorname{ran} H)$. Then for each $\alpha < \omega_1$, let $b_\alpha \in T_{H(\alpha)}$ be such that $b_\alpha <_T x$. For some $\alpha < \beta < \omega_1$, we must have $F(b_a) = F(b_\beta)$, of course. This contradiction proves the result. □

Our third application is indirect, in that it follows from the previous two. If two is due to Baumgartner, though the present proof involves ideas of Todorcevic.

<u>Theorem 5.8</u>

Assume PFA + $2^{\aleph_1} = \aleph_2$. Let $E = \{\alpha<\omega_2 \mid cf(\alpha) = \omega_1\}$. Then $\diamondsuit(E)$ holds.

Proof : We must find a sequence $\langle W_\alpha \mid \alpha \in E \rangle$ such that $W_\alpha \subseteq {}^{\alpha}2$ $|W_\alpha| \leq \aleph_1$, and for any $f \in {}^{\omega_2}2$ there is an $\alpha \in E$ such that $f \upharpoonright \alpha \in W_\alpha$.

For each $\alpha < \omega_2$, let $\langle f^\alpha_\xi \mid \xi < \omega_2 \rangle$ enumerate ${}^{\alpha}2$. For each $f \in {}^{\omega_2}2$, we can define $F_f : \omega_2 \to \omega_2$ by $F_f(\alpha) = \min\{\xi \mid f\upharpoonright\alpha = f^\alpha_\xi\}$. Similarly, if $\alpha < \omega_2$ and $f \in {}^{\alpha}2$ we define $F_f : \alpha + 1 \to \omega_2$ by $F_f(\beta) = \min\{\xi \mid f\upharpoonright\beta = f^\beta_\xi\}$.

For $\phi, \psi \in {}^{\omega_2}\mathrm{On}$, say $\phi \prec \psi$ iff $|\{\alpha \mid \phi(\alpha) \geq \psi(\alpha)\}| \leq \aleph_1$. Clearly, $\prec$ is a well-founded partial ordering of ${}^{\omega_2}\mathrm{On}$.

We define W_α, $\alpha \in E$, by recursion on α. So let $\alpha \in E$ be such that W_β is defined for all $\beta \in E \cap \alpha$.

Let $\mathfrak{J}_\alpha$ denote the set all trees T with the following properties:

(i) T is an initial segment of ${}^{\alpha}2$ of cardinality $\aleph_1$;

(ii) every element of T is contained in an α-branch of T;

(iii) if $f \in T$ then $(\forall\gamma<\alpha)(f\upharpoonright\gamma \notin W_\gamma)$;

(iv) if $\gamma < \alpha$ and $f^\gamma_\xi \in T$, then if $\zeta < \xi$ is such that $(\forall\delta<\alpha)(f^\gamma_\zeta\upharpoonright\delta \notin W_\delta)$, then $f^\gamma_\zeta \in T$.

If $\mathfrak{J}_\alpha \neq \phi$ we define $T_\alpha = \bigcap\mathfrak{J}_\alpha$. Clearly, T_α is a tree satisfying (i), (iii) and (iv). We show that T_α also satisfies (ii), whence $T_\alpha \in \mathfrak{J}_\alpha$. Let $f \in T_\alpha$. Let $T^0 \in \mathfrak{J}_\alpha$ and let b^0 be an α-branch of T^0 containing f. Set $g^0 = \bigcup b^0 \in {}^{\alpha}2$. Now, if $b^0 \subseteq T$ for all $T \in \mathfrak{J}_\alpha$ we are done. Otherwise choose $T^1 \in \mathfrak{J}_\alpha$ so that $b^0 \nsubseteq T^1$. Let b^1 be an α-branch of T^1 containing f. Set $g^1 = \bigcup b^1 \in {}^{\alpha}2$. Using properties (iii), (iv), and (i) we conclude that for some $\gamma_1 < \alpha$, $\gamma_1 \leq \delta < \alpha \to F_{g_1}(\delta) < F_{g_0}(\delta)$. [Otherwise there are arbitrarily large $\delta < \alpha$ so that $F_{g_0}(\delta) = \min\{\xi | g_0 \upharpoonright \delta = f_\xi^\delta\} \leq \min\{\xi | g_1 \upharpoonright \delta = f_\xi^\delta\} = F_{g_1}(\delta)$, so by (iii) and (iv), $g_0 \upharpoonright \delta \in T^1$ for arbitrarily large δ, so by (i), $g_0 \upharpoonright \delta \in T^1$ for all $\delta < \alpha$, giving $b^0 \subseteq T^1$, a contradiction.] If $b^1 \subseteq T$ we are done. Otherwise choose $T^2 \in \mathcal{J}_\alpha$ so that $b^1 \nsubseteq T^2$ and let b^2 be an α-branch of T^2 containing f. Set $g^2 = \cup b^2 \in {}^\alpha 2$. Then for some $\gamma_2 < \alpha$, $\gamma_2 \leq \delta < \alpha \to F_{g_2}(\delta) < F_{g_1}(\delta)$. And so on. If this process ever terminates we are done. Otherwise, if we set $\gamma = \sup_{n<\omega}\gamma_n$, then $\gamma < \alpha$, since $cf(\alpha) = \omega_1$. But then we have $F_{g_0}(\gamma) > F_{g_1}(\gamma) > F_{g_2}(\gamma) > \ldots$, a contradiction. Thus $T_\alpha \in \mathcal{J}_\alpha$.

If $\mathcal{J}_\alpha = \phi$ we set $W_\alpha = \{f_\xi^\alpha \mid \xi < \alpha\}$. Otherwise we set

$$W_\alpha = \{\cup b \mid b \text{ is an } \alpha\text{-branch of } T_\alpha\}.$$

By Theorem 5.6, in either case we have $|W_\alpha| \leq \aleph_1$. That completes the definition of $\langle W_\alpha \mid \alpha \in E\rangle$.

We show that $\langle W_\alpha \mid \alpha \in E\rangle$ is a $\Diamond(E)$-sequence. Suppose otherwise, and let $f_0 \in {}^{\omega_2}2$ be such that $(\forall\alpha\in E)(f_0 \upharpoonright \alpha \notin W_\alpha)$, but if $g \in {}^{\omega_2}2$ is such that $F_g \prec F_{f_0}$, then $(\exists\alpha\in E)(g \upharpoonright \alpha \in W_\alpha)$. Let $S = \{f \upharpoonright \alpha \mid \alpha < \omega_2 \ \& \ f \in {}^{\omega_2}2 \ \& \ (\forall\alpha\in E)(f \upharpoonright \alpha \notin W_\alpha)\}$, an initial part of ${}^{\omega_2}2$ such that every element is contained in an ω_2-branch. (By our initial assumption, $S \neq \phi$.)

Let $N \prec H_{\omega_3}$ be of cardinality $\aleph_1$ such that $\alpha = N \cap \omega_2 \in E$ and $f_0, S, \langle W_\alpha \mid \alpha \in E\rangle, \langle\langle f_\xi^\beta \mid \xi < \omega_2\rangle \mid \beta < \omega_2\rangle \in N$. Set $T = S \cap N$. Thus T is an α-subtree of ${}^\alpha 2$ of cardinality $\aleph_1$. It is easily seen that $T \in \mathcal{J}_\alpha$. Thus $T_\alpha \subseteq T \subseteq N$.

Let b be an α-branch of T_α, and set $g_0 = \bigcup b \in {}^{\alpha}2$. Now, by the choice of f_0, the α-branch determined by f_0 is not a subset of T_α. So for some $\gamma < \alpha$, $\gamma \le \delta < \alpha \to F_{g_0}(\delta) < F_{f_0}(\delta)$. [Argue as before.] Let $g_1 = g_0 \restriction \gamma$. Then $g_1 \in T_\alpha \subseteq T \subseteq N$.

Let

$$U = \{h \in {}^{\omega_2}2 \mid [h \subseteq g_1] \text{ or } [g_1 \subseteq h \ \& \ (\forall \beta \in E)(h \restriction \beta \notin W\beta) \ \& \ \& \ (\gamma \le \delta < \omega_2 \to F_h(\delta) < F_{f_0}(\delta))]\} .$$

U is an initial segment of ${}^{\omega_2}2$. Moreover, $U \in N$, since all parameters involved in the above definition are in N. Since $\gamma \le \delta < \alpha \to g_0 \restriction \delta \in U$, $N \models$ "U has height ω_2". Hence as $N \prec H_{\omega_3}$, U really does have height ω_2. Let $\gamma \le \delta < \omega_2$, and let $h \in U \cap {}^{\delta}2$. Let $\xi = \min\{\zeta \mid h = f_\zeta^\delta\}$. By definition of U, $\xi < F_{f_0}(\delta)$. Thus $|U \cap {}^{\delta}2| \le \aleph_1$ for all $\delta < \omega_2$. By Theorem 5.7, U has an ω_2-branch, say d. Let $g = \bigcup d \in {}^{\omega_2}2$. Then $F_g(\delta) < F_{f_0}(\delta)$ for all $\delta \ge \gamma$, so $F_g \prec F_{f_0}$. But $d \subseteq U$, so $(\forall \beta \in E)(g \restriction \beta \notin W_\beta)$, so this contradicts the choice of f_0. The proof is complete. ◻

For many more applications of PFA we refer the reader to [Ba].

§6. HISTORICAL REMARKS

As we said in the introduction, the theory of proper forcing was worked out by Saharon Shelah. As far as I am aware, the first appearance in print of the general notion occurred in [Sh1]. Much of the more advanced theory of proper forcing - which I have not mentioned in my present account — was developed in a series of "open letters" from Shelah to E. Wimmers [S12], photocopies of which have been distributed to "the cognoscenti". Some of this material is promised to appear in [W].

References

[Ba] J.E. Baumgartner. *Some Applications of PFA*. In preparation.

[DeJo] K.J. Devlin and H. Johnsbråten. *The Souslin Problem*. Springer Lecture Notes in Mathematics 405 (1974).

[DeSh] K.J. Devlin and S. Shelah. *A Weak Version of ◇ Which Follows from $2^{\aleph_0} < 2^{\aleph_1}$*. Israel Journal of Mathematics 29 (1978), pp. 239-247.

[Ek] P.C. Eklof. *Set Theoretical Methods in Homological Algebra and Abelian Groups*. University of Montreal Press (1980).

[Je] T.J. Jech. *Set Theory*. Academic Press (1978).

[S11] S. Shelah. *Independence Results*. JSL 45 (1980), pp.563-573.

[S12] S. Shelah. Letters to E. Wimmers.

[Sh] J.R. Shoenfield. *Unramified Forcing*. A.M.S. Proc. Sym. in Pure Math. XIII (1971), pp.357-381.

[SoTe] R.M. Solovay and S. Tennenbaum. *Iterated Cohen Extensions and Souslin's Problem*. Annals of Math. 94 (1971) pp.201-245.

[W] E. Wimmers. *The Shelah p-Point Independence Theorem*. Israel Journal of Mathematics. (To appear).

The Singular Cardinals Problem
Independence Results

Saharon Shelah
Institute of Mathematics
The Hebrew University, Jerusalem, Israel

Abstract: Assuming the consistency of a supercompact cardinal, we prove the consistency of

1) $\aleph_\omega$ strong limit, $2^{\aleph_\omega} = \aleph_{\alpha+1}$, $\alpha < \omega_1$ arbitrary;

2) $\aleph_{\omega_1}$ strong limit, $2^{\aleph_{\omega_1}} = \aleph_{\alpha+1}$, $\alpha < \omega_2$ arbitrary;

3) $\aleph_\delta$ strong limit, cf $\delta = \aleph_0$, $2^{\aleph_\delta}$ arbitrarily large before the first inaccessible cardinal; for $\aleph_\delta$ "large" enough.

Our work continues that of Magidor [Mg 1][Mg 2].

The author would like to thank the United States-Israel Binational Science Foundation for partially supporting this research by Grant 1110.

The author thanks Menachem Magidor for patiently listening to some wrong proofs and to the present one.

Notation:

Let $i,j,\alpha,\beta,\gamma,\xi,\zeta$ be ordinals, δ a limit ordinal, λ,μ,κ,χ cardinals (usually infinite) ℓ,k,n,m, natural numbers.

Let $P_{<\kappa}(A) = \{t\colon t \text{ a subset of cardinality} < \kappa\}$. We let t,s denote members of $P_{<\kappa}(\lambda)$.

Notation on forcing:

$\underline{P},Q,R$ denote forcing notions, i.e. partial orders, $R \subseteq Q$ means every element of R is an element of Q and on R the partial orders are equal. Let $R < Q$ mean $R \subseteq Q$, any two elements of R are compatible in R iff they are compatible in Q and every maximal antichain of R is a maximal antichain of Q.

Let $\mathrm{Col}(\lambda,<\kappa) = \{f\colon f$ a partial function from $(\kappa-\lambda) \times \lambda$ to κ, $f(\alpha,i) < \alpha$, f has power $< \lambda\}$.

We let π,σ denote members of $\underline{P}$, q,r members of Q or R. We say π,σ are compatible if they have a common upper bound and equivalent if they are compatible with the same members of $\underline{P}$. Note that π,σ are equivalent iff for any generic G, $\pi \in G \Leftrightarrow \sigma \in G$.

For any forcing notion, let $\emptyset$ be its minimal element.

§1.

1.1. Framework: In our universe V, κ is λ_n-supercompact for $n < \omega$, $\lambda_n \leq \lambda_{n+1}$, moreover the λ_n-supercompactness is preserved by any κ-directed complete forcing notion (see Laver [L]). R_n is a κ-complete forcing notion. $R < R_{n+1}$, $\Vdash_{R_n}$ "$\lambda_n^{<\kappa} = \lambda_n$". So if we force by R_n, κ is still λ_n-supercompact (more exactly $|\lambda_n|$-supercompact, as maybe λ_n was collapsed), so there is an R_n-name $\underset{\sim}{E}_n$ of a normal fine ultrafilter on $P_{<\kappa}(\lambda_n) = \{t\colon t$ a subset of λ_n of power $<\kappa\}$. Note that $P_{<\kappa}(\lambda_n)^{V^{R_n}}$ belongs to V, is included in it and is $P_{<\kappa}(\lambda_n)^V$; as forcing by R_n does not add sequences of ordinals of length $<\kappa$. But the members of $\underset{\sim}{E}_n$ (which are subsets of $P_{<\kappa}(\lambda_n)$) are not necessarily from V.

Let for $t \in P_{<\kappa}(\lambda_n)$, $\alpha < \lambda_n$, $\alpha(t)$ be the order type of $\alpha \cap t$.

Let $I_n \subseteq P_{<\kappa}(\lambda_n)$, $\Vdash_{R_n}$ "$I_n \in \underset{\sim}{E}_n$". $t \in I_n \Rightarrow [\lambda_n(t)^{<\kappa(t)} = \lambda_n(t)$ and $t \cap \kappa$ strongly inaccessible] possible as we have assumed $\Vdash_{R_n}$ "$\lambda_n^{<\kappa} = \lambda_n$".

Let $t \underset{\sim}{\subseteq} s$ mean $t \subseteq s$, $|t| < \kappa(s)$.

We let $\underset{\sim}{C}$ be an R_n-name for every n, i.e. $\underset{\sim}{C}_n$ is an $\underset{\sim}{R}_n$-name for every n, and $\Vdash_{R_n}$ "$\underset{\sim}{C}_n = \underset{\sim}{C}_{n+1}$".

<u>1.2 The forcing notion:</u> The forcing notion $\underline{P}$ we shall use is defined as follows:

An element π of $\underline{P}$ has the form:

$$<r, \bar{t}, \bar{f}, \bar{A}, \bar{G}>$$

where for some $n < \omega$:

A) $r \in R_0$

B) $\bar{t} = <t_1, \ldots, t_n>$, $t_\ell \in I_\ell$, $t_\ell \underset{\sim}{\subseteq} t_{\ell+1}$

C) $\bar{f} = <\underset{\sim}{f}_0, \ldots, \underset{\sim}{f}_n>$, $\underset{\sim}{f}_\ell$ an $R_{\ell+1}$-name

D) Let $\kappa_\ell = \kappa(t_\ell)$ - (which is the order-type of $\kappa \cap t_\ell$), then $\underset{\sim}{f}_0 \in \mathrm{Col}(\aleph_1, < \kappa_1)$, $\underset{\sim}{f}_\ell \in \mathrm{Col}(\lambda_n(t_n)^+, \kappa_{\ell+1})$ for $1 \leq \ell < n$ and $\underset{\sim}{f}_n \in \mathrm{Col}(\lambda_n(t_n)^+, < \kappa)$ (i.e. those things are forced, but $\underset{\sim}{f}_\ell$ is a name of an element of V, so we omit the ~ if we know the value and write $\bar{f} \in V (f_\ell \in V)$.)

E) $\bar{A} = <\underset{\sim}{A}_\ell: n < \ell < \omega>$, $\underset{\sim}{A}_\ell$ is an R_ℓ-name of a member of $\underset{\sim}{E}_\ell$

F) $\bar{G} = <G_\ell: n < \ell < \omega>$, $\underset{\sim}{G}_\ell$ is an $R_{\ell+1}$-name of a function with domain I_ℓ, and $\underset{\sim}{G}_\ell(t) \in \mathrm{Col}(\lambda_\ell(t)^+, <\kappa)$.

We write $n = n[\pi]$, $t_i = t_i[\pi]$ etc., or $n = n^\pi$, etc.

1.2(A) The order on $\underset{\sim}{P}$:

The order is natural: $\pi \le \sigma$ *iff*

A) $r^\pi \le r^\sigma$ (in R_O)

B) $n^\pi \le n^\sigma$ and $t^\pi_\ell = t^\sigma_\ell$ for $\ell = 1,\dots,n^\pi$

C) $\underset{\sim}{f}^\pi_\ell \subseteq \underset{\sim}{f}^\sigma_\ell$ for $\ell = 0,\dots,n^\pi$ (i.e. $r^\sigma \Vdash_{R_O}$ "$\underset{\sim}{f}^\pi_\ell \subseteq \underset{\sim}{f}^\sigma_\ell$")

D) $r^\sigma \Vdash_{R_O}$ "$t^\sigma_\ell \in \underset{\sim}{A}^\pi_\ell$ and $G^\pi_\ell(t^\sigma_\ell) \subseteq \underset{\sim}{f}^\sigma_\ell$ "

for $\ell = n^\pi + 1,\dots,n^\sigma$

E) $\underset{\sim}{A}^\pi_\ell \supseteq A^\sigma_\ell$, $(\forall t \in I_\ell)$ $[\underset{\sim}{G}^\pi_\ell \subseteq \underset{\sim}{G}^\sigma_\ell]$ (i.e., this is forced by r^σ)

1.2 (B) Claim: The set of $\pi \in \underset{\sim}{P}$, such that $f^\pi \in V$ is a dense subset of $\underset{\sim}{P}$. So usually we deal with such π only.

1.3 Technical Definitions on the forcing conditions:

1.3 A Definition: For $\pi,\sigma \in \underset{\sim}{P}$ we call σ a j-direct extension of π if

a) $\pi \le \sigma$

b) $\underset{\sim}{f}_\ell[\pi] = \underset{\sim}{f}_\ell[\sigma]$ for $j \le \ell \le n^\pi$

c) $q[\sigma] \Vdash_{R_O}$ "$\underset{\sim}{G}(t_\ell) = \underset{\sim}{f}_\ell[\sigma]$" for $n^\pi < \ell \le n^\sigma$

d) $\underset{\sim}{A}_\ell[\pi] = \underset{\sim}{A}_\ell[\sigma]$ for $\ell > n^\sigma$

e) $\underset{\sim}{G}_\ell[\pi] = \underset{\sim}{G}_\ell[\sigma]$ for $\ell > n^\sigma$

Convention:

We omit j when $j = n^\pi + 1$.

1.3 B Definition: For $\pi,\sigma \in \underset{\sim}{P}$ we call σ a j-length preserving extension of π if

(a) $\pi \le \sigma$

(b) $n^\pi = n^\sigma$

(c) $\underset{\sim}{f}^\pi_\ell = \underset{\sim}{f}^\sigma_\ell$ for $\ell < j$

Convention

We omit j when $j = n^{\pi} + 1$

1.3.C Definition: 1) For $\pi, \sigma \in \underline{P}$, we call σ an R-extension of π if $\pi \leq \sigma$, $\bar{t}^{\pi} = \bar{t}^{\sigma}, \bar{f}^{\pi} = \bar{f}^{\sigma}$, $\underset{\sim}{A}^{\pi} = \underset{\sim}{A}^{\sigma}$, $\underset{\sim}{G}^{\pi} = \underset{\sim}{G}^{\sigma}$, or at least if r^{σ} forces those inequalities.

2) For $\pi, \sigma \in \underline{P}$ we call σ an R-constant extension of π if $\pi \leq \sigma$, $r^{\pi} = r^{\sigma}$.

1.3.D Claim and Definition:

If $j < \omega$, $\pi_1 \leq \pi_2$ then there is a unique π such that π is a j-direct extension of π_1, π_2 is a j-length preserving extension of π, and $r[\pi_1] = r[\pi_2]$ [and $r[\pi] = r[\pi_1]$].

This unique π is called the upper [lower] j-interpolant of π_1, π_2. If $j = n[\pi_2] + 1$ we omit j.

1.4 The Inner Model:

The forcing $\underline{P}$ gives too much, e.g. it collapses all cardinals which are both $\leq \sum_n \lambda_n$ and $> \kappa$, and maybe also κ. But we shall use an inner model. Define some $\underline{P}$-names: $\underset{\sim}{t}_n, \underset{\sim}{\kappa}_n$ are $t_n^{\pi}, \kappa(t_n^{\pi})$ (for every large enough π in the generic set), $\underset{\sim}{F}_{\ell} = \cup \{f_{\ell}^{\pi}$: π in the generic set$\}$, and $\underset{\sim}{C}$ as an R_0-name is a $\underline{P}$-name.

For a generic $G \subseteq \underline{P}$, we shall be interssted in the inner model $V[\langle\kappa_{\ell}[G], \underset{\sim}{F}_{\ell}[G] : \ell < \omega\rangle, \underset{\sim}{C}[G]]$; let $\underset{\sim}{V}_f$ be a P-name of this class.

1.5 Automorphism of $\underline{P}$:

The proofs of the following are well known.

<u>1.5 A Claim</u>: Suppose H is an automorphism of $\underline{P}$, then it induces naturally a permutation of the set of $\underline{P}$-names $a \to \underset{\sim}{a}^H$ and if $\underset{\sim}{a}_1,\ldots,\underset{\sim}{a}_n$ are $\underline{P}$-names, $\phi(x_1,\ldots,x_n)$ a first order formula, $\pi \in \underline{P}$, <u>then</u> $\pi \Vdash$ "$\phi(\underset{\sim}{a}_1,\ldots,\underset{\sim}{a}_n)$" iff $H(\pi) \Vdash_{\underline{P}}$ "$\phi(\underset{\sim}{a}_1^H,\ldots,\underset{\sim}{a}_n^H)$". We say that H preserves a if $\Vdash_{\underline{P}}$ "$\underset{\sim}{a} = \underset{\sim}{a}^H$".

<u>1.5 B Claim</u>: If $\pi \Vdash_P$ "$\underset{\sim}{a} \in \underset{\sim}{V}_f$" then there is a P-name b, such that $\pi \Vdash_P$ "$\underset{\sim}{a} = \underset{\sim}{b}$", and every automorphism of $\underline{P}$ which preserves $\underset{\sim}{V}_f$ (i.e. preserve $\underset{\sim}{\kappa}_\ell$, $\underset{\sim}{f}_\ell$, $\underset{\sim}{C}$) preserves also $\underset{\sim}{b}$.

<u>1.5 C Claim</u>: Let H be a permutation of $\bigcup_{n<\omega} \lambda_n$, which maps $\lambda_n (n < \omega)$ onto themselves, $H \restriction \kappa$ = the identity:

1) H induces an automorphism of $\underline{P}$, which we denote by H too, as follows:

$$\text{for } t \in I_n (n < \omega) \text{ let } H(t) = \{H(i): i \in t\}$$

$$r^{H(\pi)} = r^\pi$$

$$\bar{t}^{H(\pi)} = \langle H(t_\ell^\pi): 1 \le \ell \le n^\pi \rangle$$

$$\bar{f}^{H(\pi)} = \bar{f}$$

$$\underset{\sim}{A}_\ell^{H(\pi)} = \{H^{-1}(t): t \in \underset{\sim}{A}_\ell\}$$

$$\underset{\sim}{G}_\ell^{H(\pi)}(t) = \underset{\sim}{G}_\ell^\pi(H^{-1}(t))$$

2) Note also that $\Vdash_{R_n}$ "$\{t \in I_n: H(t) = t\} \in \underset{\sim}{E}_n$" and that H preserves $\underset{\sim}{V}_f$.

<u>1.5 D Definition</u>: We call $\underset{\sim}{a}$ a $\underset{\sim}{V}_f$-name if it is a $\underline{P}$-name and is preserved by any automorphism of $\underline{P}$ preserving $\underset{\sim}{V}_f$, and $\Vdash_{\underline{P}}$ "$\underset{\sim}{a} \in \underset{\sim}{V}_f$".

<u>1.5 E Claim</u>: Suppose H is a permutation of $\bigcup\lambda_n$, mapping each λ_n onto itself, $H \restriction \kappa$ = the identity.

Then for any $\pi \in \underset{\sim}{P}$, $H(\pi)$ and $H^*(\pi) = \langle r^\pi, \bar{t}^{H(\pi)}, \bar{f}^{H(\pi)}, \bar{A}^\pi, G^\pi \rangle$ are compatible. Moreover suitable increasing of the $\underset{\sim}{\bar{A}}$ makes them equivalent.

Proof: By 1.5 C(2), remembering that the $\underset{\sim}{E}_\ell$'s are ultrafilters.

1.6 Claim:

Suppose a) $\pi, \sigma \in \underset{\sim}{P}$, $n(\pi) = n(\sigma) = n$, $\bar{t}^\pi = \bar{t}^\sigma, \bar{f}^\pi = \bar{f}^\sigma$ and is in V, $\underset{\sim}{A}^\pi = \underset{\sim}{A}^\sigma$ and $\underset{\sim}{G}^\pi = \underset{\sim}{G}^\sigma$.

b) every $r \in R_n$ compatible with $r[\sigma]$ is compatible with $r[\pi]$

c) $\underset{\sim}{a}$ is a $\underset{\sim}{V}_f$-name.

Then if $\pi \Vdash_P$ "$\underset{\sim}{a} = a$" (for some $a \in V$) then $\sigma \Vdash_{\underset{\sim}{P}}$ "$\underset{\sim}{a} = a$".

Proof: Easy.

1.7 Definition. Good Cardinals for P

A cardinal μ is good for R_n or for n in short, if $\mu = \mu^{<\kappa}$ and there are forcing notions Q^0_m, Q^1_m, $R_m = Q^0_m * \underset{\sim}{Q}^1_m$, Q^0_m is μ^+-complete and $\underset{\sim}{Q}^1_m$ satisfies the μ^+-chain condition (i.e. $\Vdash_{Q^0_m}$ "$\underset{\sim}{Q}^1_m$ satisfies the μ^+-chain condition").

Remark: 1) Really $Q^0_m * \underset{\sim}{Q}^1_m > R_m$ is sufficient.

2) Note that $\underset{\sim}{Q}^1_m$ is not required to be κ-complete, so $R_m = Q^0_m \times Q^1_m$, Q^0_m μ^+-complete, Q^1_m satisfying the μ^+-chain condition is sufficient for this definition.

1.8 The Main Lemma:

A) If κ is good for every $n \geq n_0$, then in V_f κ is strong limit. Moreover every subset of $\lambda_n(\underset{\sim}{t}_n)$ belong to $L[\langle F_\ell : \ell \leq n \rangle]$, hence $2^{\lambda_n(t_n)} = \lambda_n(t_n)^+$ (the + is in V_f and in V too).

B) If $\mu \geq \kappa$ is good for every $n \geq n_0$, $\Vdash_{R_n}$ "μ is regular", then in V_f μ^+ is still a regular cardinal. Moreover for any function g from μ to ordinals, for some $A \in V$, $V \models$ "$|A| = \mu$" and Range(g) $\subseteq A$.

The proof is broken to a series of claims.

<u>1.9. Notation:</u>

For $m < \omega$, let $\underset{\sim}{P}_m^{nt} = \{\langle \bar{t}^\pi, \bar{f}^\pi \rangle : \pi \in \underset{\sim}{P},\ \bar{f}^\pi \in V \text{ and } n^\pi = m\}$. For any μ, $\kappa \leq \mu \leq \lambda$ we define an equivalence relation $\approx_\mu^m$ on $\underset{\sim}{P}_m^{nt}$: $\langle \bar{t}^1, \bar{f}^1 \rangle \approx_\mu^m \langle \bar{t}^2, \bar{f}^2 \rangle$ iff $t_\ell^1 \cap \mu = t_\ell^2 \cap \mu$, $f_\ell^1 = f_\ell^2$, and there is a permutation of λ_m which is the identity on μ, preserves $\kappa, \lambda_0, \ldots, \lambda_m$ and maps t_ℓ^1 onto t_ℓ^2 (all for $1 \leq \ell \leq m$).

<u>1.9 A Claim:</u> $\approx_\mu^m$ has $\mu^{<\kappa}$ equivalence classes, and we can find $\langle \bar{t}^{i,j}, \bar{f}^{i,j} \rangle$ ($i < \mu^{<\kappa}, j < \mu^+$) such that: $\langle t^{i,j}, \bar{f}^{i,j} \rangle / \approx_\mu^m$ depend on i only, every $\approx_\mu^m$-equivalence class is represented by some $\langle \bar{t}^{i,j}, \bar{f}^{i,j} \rangle$, and if $\lambda_m > \mu$, there are $\alpha_{i,j} < \lambda_m$ defined when $t_m^{i,j} \not\subseteq \mu$ which belong to $t_m^{i',j'}$ iff $i = i'$, $j = j'$. We can assume moreover $t_m^{i,j} \cap t_m^{i(1),j(1)} \subseteq \mu$.

<u>1.10. Claim:</u>

Suppose $\pi \in \underset{\sim}{P}$, $m < \omega$, g a $\underset{\sim}{V}_f$-name of a function from μ to ordinals, $\pi \Vdash_{\underset{\sim}{P}}$ "$\underset{\sim}{t}_m \not\subseteq \mu$". Suppose further that μ is good for m and $\mu = \mu^{<\kappa}$.

Then there is a length preserving extension σ of π, such that if $\sigma \leq \sigma' \in \underset{\sim}{P}$, $i < \mu$, $\sigma' \Vdash_{\underset{\sim}{P}}$ "$g(i) = \sigma$", $m\lceil\sigma'\rceil = m$ then also the upper interpolant σ^{up} of σ and σ' force this. Moreover, for some set A of ordinals, $|A| \leq \mu$ ($A \in V$ of course), for every

R-extension σ^* of the lower interpolant of σ and σ', if $\sigma^* \Vdash_{\underline{P}}$ "$g(i) = \alpha$", then $\alpha \in A$.

Proof of Claim 1.10:

So $R_m = Q^0_m * \underset{\sim}{Q}^1_m$, and $R_0 < R_m$, hence there is a pair $\langle q^0, \underset{\sim}{q}^1 \rangle \in Q^0_m * \underset{\sim}{Q}^1_m$, such that every extension of it in $Q^0_m * \underset{\sim}{Q}^1_m$ is compatible with r^π (which belong to R_0).

We now define for $i < \mu$, ordinals $\alpha_i < \mu^+$, and for $j < \alpha_i$ a condition $\pi_{i,j}$, such that

A) $r[\pi_{i,j}] = \langle q^0_{i,j}, \underset{\sim}{q}^1_{i,j} \rangle$, $q^0_{i,j} \le q^0_{\xi,\zeta}$ when $\langle i,j \rangle < \langle \xi,\zeta \rangle$ (i.e $i < \xi$ or $i = \xi$, $j < \zeta$).

B) for each i, $\{q^1_{i,j+1} : j+1 < \alpha_i\}$ is a maximal antichain (of $\underset{\sim}{Q}^1_m$) (i.e. $q^0_{i+1,0}$ forced this).

C) $\bar{t}[\pi_{i,j+1}] = \bar{t}^{i,j}$, $\bar{f}[\pi_{i,j+1}] = \bar{f}^{i,j}$, $n[\pi_{i,j}] = m$ (see Claim 1.9A).

D) $\pi_{0,0}$ is a direct extension of π.

E) For each $\alpha \in t^{i,j}_m$ either $\pi_{i,j+1}$ determines (i.e. forces) a value for $\underset{\sim}{g}(\alpha)$ or there is no length preserving $\pi' \ge \pi_{i,j}$ which does so.

F) For $\ell > m$, $\Vdash_{R_\ell}$ $(\forall t \in \underset{\sim}{A}_t[\pi_{i,j+1}])(t^{i,j}_m \underset{\sim}{\subseteq} t)$.

G) For $\ell > m$, $\underset{\sim}{G}_\ell[\pi_{i,j}]$ increases, i.e. if $\langle i,j \rangle \le \langle \xi,\zeta \rangle$, $t \in I_\ell$ then $\Vdash_{R_\ell}$ " if (1) $t \in \underset{\sim}{A}_\ell[\pi_{i,j}]$ and $t \in \underset{\sim}{A}_\ell[\pi_{\xi,\zeta}]$ or (2) $i = j = 0$ then $\underset{\sim}{G}_\ell[\pi_{i,j}](t) \subseteq \underset{\sim}{G}_\ell[\pi_{\xi,\zeta}](t)$".

H) $\underset{\sim}{G}_\ell[\pi_{i,j}]$ does not increase unnecessarily, i.e. $\Vdash_{R_\ell}$ "if $t \notin \underset{\sim}{A}_\ell[\pi_{i,j}]$ or j is not a successor or $\underset{\sim}{q}^1_{i,j}$ is not in the generic set then $G_\ell[\pi_{i,j}](t)$ is the union of $G_\ell[\pi_{\xi,\zeta}]$ $\langle \xi\ \zeta \rangle < \langle i,j \rangle$, $t \in \underset{\sim}{A}[\pi_{\xi,\zeta}]$ or $i = j = 0$".

Note that $\underset{\sim}{G}_\ell[\pi_{i,j}](t)$ is increased only when $\underset{\sim}{t}_m^{i,j} \subseteq t$ so this occurs $\le |t|^{<\kappa(t)} = |t| = \lambda_\ell(t)$ times, but $col(\lambda_\ell(t)^+, <\kappa)$ is $\lambda_\ell(t)^+$-complete, so we can continue to define. So there is no problem to carry the construction by induction on $<i,j>$ (the α_i's are defined as Q_m^1 satisfies the μ^+-chain condition). In the end we have to define σ. For $r[\sigma]$, note first that $\{q^0_{i,j} : i < \mu, j < \alpha_i\}$ has an upper bound $q^0_\mu \in Q^0_\mu$ as Q^0_μ is μ^+-complete (and by A)), and $r[\pi]$, $(q^0_\mu, \underset{\sim}{q}^1)$ are compatible by the choice of $(q^0, \underset{\sim}{q}^1)$, and let $r[\sigma]$ be any upper bound of $r, (q^0_\mu, \underset{\sim}{q}^1)$. Obviously, $\bar{t}^\sigma = \bar{t}^\pi$, $\bar{f}^\sigma = \bar{f}^\pi$. Now $\underset{\sim}{A}^\sigma_\ell = \{t \in I_\ell : t \in \underset{\approx}{A}_\ell(\pi)$ and for any $i < \mu$, $j < \alpha_i$, $t^{i,j} \underset{\sim}{\subseteq} t$ implies $t \in \underset{\sim}{A}_\ell[\pi_{i,j+1}]$. Moreover if $s \underset{\sim}{\subseteq} t$, $s \approx^m_\mu t^{i,j}$, $f^{i,j}_m \subseteq \kappa(t) \times \kappa(t)$, H a permutation of $\bigcup_i \lambda_i$ which is the identity except interchanging $t^{i,j}$ and s then $t = H(t) \in H[\underset{\sim}{A}_\ell[\pi_{i,j}]]$.$\}$ $\underset{\sim}{A}_\ell$ is an R_ℓ-name as each $\underset{\sim}{A}_\ell[\pi_{i,j}]$ is, and is forced to be in $\underset{\sim}{E}_\ell$ by the normality of $\underset{\sim}{E}_\ell$ (a conclusion of it, more exactly).

Now $\underset{\sim}{G}_\ell[\sigma](t)$ is the union of $G_\ell[\pi_{i,j}](t)$, $t \in \underset{\sim}{A}[\pi_{i,j}]$ or $i = j = 0$. It is easy to check everything, because $\underset{\sim}{g} \in \underset{\sim}{V}_f$, (and use 5D, 6).

<u>1.11. Claim:</u>

Suppose $\pi \in \underline{P}$, $\underset{\sim}{g}$ a $\underset{\sim}{V}_f$-name of a function from π to ordinals, $\pi \Vdash_{\underline{P}}$ "for every $m > n^\pi$, $\underset{\sim}{t}_m \not\subseteq \mu$" and $\mu = \mu^{<\kappa}$.

Then there is a length preserving extension σ of π, such that if $\sigma \le \sigma' \in \underline{P}$, μ good for $n[\sigma']$, $i < \mu$, $i \in t_{n[\sigma']}[\sigma']$, and $\sigma' \Vdash_{\underline{P}}$ "$\underset{\sim}{g}(i) = \alpha$" then also the upper interpolant σ'' of σ and σ'

forces this. Also if for some $\langle A_i : i < \mu\rangle \in V$, $|A_i| \le \mu$, and $\sigma \le \sigma' \in \underline{P}$, μ is good for $R_{n[\sigma']}$, $i \in \mu \cap t_{n[\sigma']}[\sigma']$, $\sigma' \Vdash_{\underline{P}}$ "$\underset{\sim}{g}(i) = \alpha$" then $\alpha \underset{\sim}{\in} A_i$. Hence if μ is good for arbitrarily large m, if $\sigma \le \sigma' \in \underline{P}, \sigma' \Vdash_{\underline{P}}$ "$\underset{\sim}{g}(i) = \alpha$" then for some direct extension σ'' of σ', the upper interpolant σ''' of σ, σ'' forces this.

Proof: Repeat claim 1.10 ω times.

1.12 Proof of the Main Lemma 1.8B.

Quite easy from Claim 11, because $\Vdash_{\underline{P}}$ "$\mu \subseteq \bigcup_{n<\omega} \underset{\sim}{t}_n$" and $\{\pi : \Vdash_{\underline{P}}$ "for every $m > n^\pi$, $\underset{\sim}{t}_m \not\subseteq \mu$"$\}$ is a dense subset of $\underline{P}$.

1.13 Claim: Suppose $\pi \in \underline{P}$, $m > n^\pi$, $\Vdash_{\underline{P}}$ "$\underset{\sim}{g} \in \underset{\sim}{V}_f$, $\underset{\sim}{g}$ a function from μ to ordinals".

Then there is $\sigma \in \underline{P}$, $\pi \le \sigma$ such that

A) π and σ are identical except that possibly $\underset{\sim}{G}_m[\sigma]$ is not equal to $\underset{\sim}{G}_m[\pi]$,

B) Suppose $\sigma_1 \in P$, $n[\sigma_1] = m$, $r[\sigma_1] \ge r[\sigma]$, $t_\ell \in I_\ell$ for $1 \le \ell \le m$, $r[\sigma_1] \Vdash_{R_0}$ "$\underset{\sim}{G}_m(t_m) = f_m[\sigma_1]$" and $\underset{\sim}{G}_\ell[\sigma_1] = \underset{\sim}{G}_\ell[\sigma]$, $\underset{\sim}{A}_\ell[\sigma_1] = \underset{\sim}{A}_\ell[\sigma]$ for $m < \ell < \omega$.

Suppose further σ_2 is an (m-1)-length preserving extension of σ_1, $n[\sigma_2] = n[\sigma_1] = m$, $\bar{t}[\sigma_2] = \bar{t}[\sigma_1]$, $f_\ell[\sigma_2] = f_\ell[\sigma_1]$ for $\ell < m$, $\bar{\underset{\sim}{A}}[\sigma_2] = \bar{\underset{\sim}{A}}[\sigma_1]$, $\bar{\underset{\sim}{G}}[\sigma_2] = \bar{\underset{\sim}{G}}[\sigma_1]$, $i \in t_m \cap \mu$, $\sigma_2 \Vdash_{\underline{P}}$ "$g(i) = \alpha$". Then $\langle r[\sigma_2], \bar{t}[\sigma_1], \bar{f}[\sigma_1], \bar{\underset{\sim}{A}}[\sigma_1], \bar{\underset{\sim}{G}}[\sigma_1]\rangle$ forces this too.

Proof: Just note that for each t_m the number of $\langle t_\ell : 1 \le \ell < m\rangle$,

$\langle f_\ell : \ell < m\rangle$ we have to consider is $\le |t_m|^{<\kappa(t_m)} = |t_m| \equiv \lambda_m(t_m)$ whereas $\mathrm{Col}(\lambda_m(t_m)^+, <\kappa)$ is $\lambda_m(t_m)^+$-complete.

Note also that we in fact, are interested in R_{m+1} only (not R_0).

1.13A Claim: The parallel claim to 13 holds for $m = n^\pi$.

1.14. Corollary. Suppose $\underset{\sim}{g}$ is a $\underset{\sim}{V}_f$-name, μ good for every $m \ge n[\pi]$, $\underset{\sim}{g}$ a function from μ to ordinals. Then there is π_1, a length-preserving extension of π such that:

if $\pi_1 \le \sigma \in P$, $i \in t_m[\sigma] \cap \mu$, $\sigma \Vdash_{\underline{P}}$ "$\underset{\sim}{g}(i) = \alpha$" *then* also $\sigma^* \Vdash_{\underline{P}}$ "$\underset{\sim}{g}(i) = \alpha$" where σ^* is defined by

$$r[\sigma^*] = r[\sigma]$$

$$n[\sigma^*] = n[\sigma],\ t_\ell[\sigma^*] = t_\ell[\sigma] \text{ for } \ell = 1,\dots,n[\sigma]$$

$$f_\ell[\sigma^*] = f_\ell[\sigma] \text{ for } \ell = 0,\dots,m-1$$

$$f_\ell[\sigma^*] = \underset{\sim}{G}_\ell[\pi_1](t_\ell) \text{ for } \ell = m,\dots,n[\sigma]$$

$$\underset{\sim}{A}_\ell[\sigma^*] = \underset{\sim}{A}_\ell[\pi_1] \text{ for } n[\sigma] < \ell < \omega$$

$$\underset{\sim}{G}_\ell[\sigma^*] = \underset{\sim}{G}_\ell[\pi_1] \text{ for } n[\sigma] < \ell < \omega.$$

Proof: Use Claim 1.11 and then 1.13 for all m.

1.15 Claim: Suppose π_1, g, $n = n[\pi_1]$ are as in corollary 1.14, $\mu \le \lambda(t_n^\pi)$ and $m < n[\pi_1]$, and $\underset{\sim}{g}$ is a function from μ to $\{0,1\}$ (i.e. $\underset{\sim}{g} \in V[\langle \kappa_\ell[G], f_\ell[G] : n \le \ell < \omega\rangle, C]$, and note that we can get $\underset{\sim}{V}_f[G]$ from this universe by forcing by the product of n Lévy collapsing).

Then there is $\pi_2 \in \underline{P}$ such that:

A) $\pi_1 \le \pi_2$, $\bar{t}[\pi_1] = t[\pi_2]$, $\bar{\underset{\sim}{G}}[\pi_1] = \bar{\underset{\sim}{G}}[\pi_2]$

B) If σ is an m-direct extension of π_2, $\alpha < \mu$, $i < 2$

$\sigma \Vdash_{\underline{P}}$ "$g(\alpha) = i$" then also π_2 forces this.

<u>Proof</u>: Fix α.

Let $W_k = \{<\bar{t},\bar{f}>$: for some m-direct extension σ of π_2, $n[\sigma] = k$,

$\bar{t}^\pi = \bar{t}$, $f^\pi \restriction k = \bar{f} \in V\}$.

For every $w = (\bar{t},\bar{f}) \in W_k$ we define an R_{k+1}-name $\underset{\sim}{V}_k(\bar{t},\bar{f})$ of an ordinal < 3: for $r \in R_k$

$r \Vdash$ "$\underset{\sim}{V}_k(\bar{t},\bar{f}) = \ell$" <u>iff</u>

A) $\ell < 2$, $<r, \bar{t},\bar{f}^\wedge<\underset{\sim}{G}_k(t_k)>, \underset{\sim}{\bar{A}}\restriction[k+1,\omega), \underset{\sim}{\bar{G}}\restriction[k+1],\omega) > \Vdash_{\underline{P}}$"$\underset{\sim}{g}(\alpha) = \ell$"

or

B) $\ell = 2$, and for no r', $r \le r' \in R_{k+1}$, $\ell' < 2$

$<r',\bar{t},\bar{f}^\wedge < \underset{\sim}{G}_k(t_k)>, \underset{\sim}{\bar{A}}\restriction[k+1,\omega), \underset{\sim}{\bar{G}}\restriction[k+1,\omega)> \Vdash_{\underline{P}}$ "$g(\alpha) = \ell$".

Now for every $k < \omega$ we define by downward induction on $\ell \le k$, for every $(\bar{t},\bar{f}) \in W_\ell$ an $R_{\ell+1}$-name $\underset{\sim}{V}_k(\bar{t},\bar{f})$ of an ordinal < 3 and $\underset{\sim}{B}_\ell(\bar{t},\bar{f})$ of a member of $\underset{\sim}{E}_{\ell+1}$.

For $\ell = k$ we have defined $\underset{\sim}{V}_k(\bar{t},\bar{f})$, and let $\underset{\sim}{B}_k(\bar{t},\bar{f}) = I_{k+1}$.

Now let $(\bar{t},\bar{f}) \in W_\ell$, $\ell < k$, $\underset{\sim}{V}_k(\bar{t}',\bar{f}')$ is defined for $(\bar{t}',\bar{f}') \in W_{\ell+1}$. Then $\underset{\sim}{V}_k(\bar{t},\bar{f})$ is the unique $i < 3$ such that:

$(\bar{t} = <t_1,\ldots,t_\ell>)$

$$\{t_{\ell+1} \in I_{\ell+1}: i = \underset{\sim}{V}_k(\bar{t}^\wedge <t_{\ell+1}>, \bar{f}^\wedge <\underset{\sim}{G}_\ell(t_\ell)>)\} \in \underset{\sim}{E}_{\ell+1}$$

(note all the names in the expression above are $\underset{\sim}{R}_{\ell+1}$-names)

and

$$\underset{\sim}{B}_k(\bar{t},\bar{f}) = \{t_{\ell+1} \in I_{\ell+1}: \underset{\sim}{V}_k(\bar{t},\bar{f}) = \underset{\sim}{V}_k(\bar{t}^\wedge <t_{\ell+1}>, \bar{f}^\wedge<\underset{\sim}{G}_\ell(t_\ell)>)\}$$

Now we can define $\underset{\sim}{A}_\ell[\pi_2]$:

$\underset{\sim}{A}_\ell[\pi_2] = \{t \in I_k$: for every $(\bar{t},\bar{f}) \in W_{\ell-1}$, such that $t_{\ell-1} \underset{\sim}{\subseteq} t$, and $k \ge \ell$, $t \in B_k(\bar{t},\bar{f})$, and of course $t \in \underset{\sim}{A}_\ell[\pi_1]\}$

This is fine for α, and as there are "few" (i.e. μ) α's there is no problem to prove the claim.

§2. Applications

For this section we make the following hypothesis.

2.1 Hypothesis: There is a universe V satisfying ZFC + G.C.H., in which κ is supercompact, moreover the supercompactness is preserved by κ-directed complete forcing.

Remark: We can weaken "κ is supercompact" by "κ is λ-supercompact" for λ suitable for each theorem, bur as long as we cannot get inner models with supercompact this is not so interesting, and anyhow clearly we get by our proof the expected results (or almost, replacing λ by λ^+).

Similarly for assuming G.C.H. - it is expected that violating C.G.H. is "harder" so we do not lose generality, and so though it seemed that we can get rid of it, there is no point in doing this.

Notation: $(\aleph_\alpha)^{+i} = \aleph_{\alpha+i}$ so $\lambda^{+1} = \lambda^+$.

2.2. Theorem: 1) For any $\alpha < \aleph_1$, there is an extension V_f of V in which κ is $\aleph_\omega$, is strong and $2^{\aleph_\omega} = \aleph_{\alpha+1}$.

2) Moreover in V_f there are $f_i \in \prod_{n<\omega} \aleph_n$ for $i < \aleph_{\alpha+1}$, such that for $i < j$, $f_i <^* f_j$, i.e., $\{n: f_i(n) < f_j(n)\}$ is co-finite.

Proof: 1) For $\alpha < \omega$ this is done in Magidor [Mg 1], so let $\alpha = \delta + k$, δ a limit ordinal, let $\delta = \bigcup_{n<\omega} D_n$, D_n finite, increasing. Let Q be any κ-complete forcing adding $\kappa^{+\alpha+2}$ subsets to κ and satisfying the κ^+-chain condition e.g. $Q = \{f$: f a partial function from $\kappa^{+\alpha+2}$ to $\{0,1\}$ of power $< \kappa\}$.

Let $T_n = \{(\beta,\gamma): 0 \le \beta < \gamma < \delta$ and $(\forall\xi \in D_n) \neg (\beta \le \xi \le \gamma)\}$

$$R_n = (\prod_{(\beta,\gamma)\in T_n} \mathrm{Col}(\kappa^{+\beta}, < \kappa^{+\gamma})) \times Q.$$

$$\lambda_n = \kappa^{+\alpha+2}.$$

Now clearly:

Fact A: R_n is κ-directed complete, $\Vdash_{R_n} 2^{\kappa} < \kappa^{+\omega}$, hence for every $t \in I_n$, $\lambda_n(t) < \kappa(t)^{+\omega}$.

Fact B: $\kappa^{+\beta}$ is good for R_n if for some γ, $\beta = \gamma + 2$, $\gamma + 1 \in D_n$ or $\beta > \delta + 1$, hence every $\gamma + 2 \le \alpha + 1$ is good for R_n for every n large enough.

Fact C: In V_f (from §1) κ is $\aleph_\omega$, $2^{\kappa_n} \le \chi_n$, $2^{\chi_n} = \lambda_n^+$ and $\kappa_n < \kappa$, where $\chi_n = \lambda_n(\underset{\sim}{t}_n)$ so κ is $\aleph_\omega$ and strong limit.

Now the theorem is immediate.

2) Just change Q to add $f_i:\kappa \to \kappa$, $(i < \kappa^{+\alpha+2})$, such that $f_i <^* f_j$ for $i < j$.

2.3. Theorem: 1) For any $\alpha < \aleph_2$ there is an extension V_f of V in which κ is $\aleph_{\omega_1}$, and is strong limit, and $2^{\aleph_{\omega_1}} = \aleph_{\alpha+1}$.

2) For any fixed $\xi < \omega_1$ we can assume that

$H(\aleph_\xi)^V = H(\aleph_\xi)^{V_f}$ ($H(\lambda)$ is the family of set of hereditary cardinality $< \lambda$).

Proof: 1) Just amalgamate the proof of 2.2 and of Magidor [Mg 1] §5 (see [Mg 3] too).

2) Just let in §1 $\underset{\sim}{f}_0 \in \mathrm{Col}(\aleph_\xi, < \kappa_1(t_1))$.

Remark: By Magidor [Mg 4] (improving Galvin and Hajnal) if Chang's conjecture holds, $\aleph_{\omega_1}$ is strong limit, then $2^{\aleph_{\omega_1}} < \aleph_{\omega_2}$. So using a method which gives 2.3(2), we cannot improve 2.3(1).

Remark: By [Sh 2], we cannot improve 2.2(1) result for $\alpha > (2^{\aleph_0})^+$ if the method gives 2.2(2) too.

2.4. Definition: 1) For a monotonic function f from ordinals to ordinals we define a function $f^{[i]}$ from ordinals to ordinals by induction on i:

$$f^{[0]}(\alpha) = \alpha$$

$$f^{[\alpha+1]}(\alpha) = f(f^{[i]}(\alpha) + 1)$$

$$f^{[\delta]}(\alpha) = \bigcup_{i<\delta} f^{[i]}(\alpha)$$

2) For a function f from ordinals to ordinals we define a function f^* from ordinals to ordinals

$$f^*(\alpha) = f^{[\alpha]}(0)$$

3) For a class C of ordinals we define by induction on α a function Suc_C^α from ordinals

$$\mathrm{Suc}_C^0(i) = \min\{\xi: \xi \in C, \xi > i\}$$

$$\mathrm{Suc}_C^{\alpha+1}(i) = (\mathrm{Suc}_C^\alpha)^*(i)$$

$$\mathrm{Suc}_C^\delta(i) = \bigcup_{\alpha<\delta} \mathrm{Suc}_C^\alpha(i)$$

4) In 3) if C is the class of infinite cardinals, then we omit it.

2.5. Lemma: Suppose λ has cofinality $\aleph_0$, and for $\chi < \lambda$, $n < \omega$, $\mathrm{Suc}^n(\chi) < \lambda$. Suppose further $\mu \geq \lambda$ but there is no weakly inaccessible cardinal κ, $\lambda \le \kappa \le \mu$. Then there are D_n, $D_n \subseteq D_{n+1}$, $\cup D_n \supseteq \{\chi: \lambda \le \chi \le \mu, \chi \text{ a cardinal}\}$, and $\mathrm{Suc}^n_{D_n}(\kappa) \ge \mu$.

Proof: We prove, by induction on μ, the existence of $\langle D_n(\mu): n < \omega \rangle$ as required.

Case I: $\mu = \lambda$.

We let $D_n(\mu) = \{\lambda\}$, and there is no problem.

Case II: $\mu = \chi^+$ for some χ, or even $\chi < \mu \le \beth_\chi$.

Let $D^* = \{\kappa: \chi < \kappa\}$, $D_0(\mu) = \{\kappa: k \ge \mu\}$, $D_1(\mu) = D^*$ $D_{n+2}(\mu) = D_n(\chi) \cup D^*$.

So $D_n(\mu)(n < \omega)$ is increasing, with union $\supseteq \{\kappa: \lambda \le \kappa \le \mu\}$; $\mathrm{Suc}^n_{D_n(\chi)}(\lambda) \ge \chi$, hence $\mathrm{Suc}^{n+2}_{D_n(\chi)}(\lambda) \ge \beth_\chi$ hence is $\ge \mu$, so $\mathrm{Suc}^n_{D_n(\mu)}(\lambda) \ge \mu$ for $n > 1$; for $n = 0,1$ this is true too by the definition.

Case III: μ is a limit cardinal, but for $\chi < \mu$, $\beth_\chi < \mu$.

Let $\mu = \sum_{i<\chi} \mu_i$, $\chi = \mathrm{cf}\mu$, $\mu_i < \mu$, μ_i increasing continuous, $\mu_0 = \beth_0$.

Let $D_0(\mu) = D_1(\mu) = \{\kappa: \kappa \ge \mu\}$, $D_{n+2}(\mu) = \{\chi$: for some i, $\mu_i \le \chi < \mu_{i+1}$, $\chi \in D_n(\mu_{i+1})$ and $\beth_i \in D_n(\beth_\chi)\}$. The checking is easy.

2.6. Theorem: 1) For any $\mu > \kappa$, $\mu^\kappa = \mu$, smaller than the first inaccessible cardinal $> \kappa$, there is a forcing extension V_f of V, in which κ is $\mathrm{Suc}^\omega(\beth_0)$, is strong limit, $2^\kappa = \mu$, and no cardinal in $[\kappa,\mu]$ is collapsed, except possibly successors of singular cardinals.

2) Moreover we can assume that in V_f there are functions $f_i \in \prod_{n<\omega} \kappa_n$ ($i < \mu$ where $\kappa = \sum_{n<\omega} \kappa_n$, $\kappa_n < \kappa_{n+1}$) such that for $i < j$, $f_i <^* f_j$.

Proof: Similar to the proof of 2.2., using lemma 2.5 (so for $t \in I_n$, $\mathrm{Suc}^n(\kappa(t)) \geq \lambda(t)$).

REFERENCES

[GH] F. Galvin and A. Hajnal, Inequalities for cardinal powers, Annals of Math., 101 (1975), 491-498.

[L] R. Laver, Making the supercompactness of κ indestructible under κ-directed closed forcing, Israel J. Math., 29 (1978), 385-388.

[Mg 1] M. Magidor, On the singular cardinals problem I, Israel J. Math., 28 (1977), 1-31.

[Mg 2] M. Magidor, On the singular cardinals problem II, Annals of Math., 106 (1977), 517-549.

[Mg 3] M. Magidor, Changing cofinality of cardinals. Fund. Math. XCIX (1978), 61-71.

[Mg 4] M. Magidor, Chang conjecture and powers of singular cardinals, J. Symb. Logic, 42 (1977), 272-276.

[Sh 1] S. Shelah, A note on cardinal exponentiation. J. Symb. Logic. 45 (1980), 56-66.

[Sh 2] S. Shelah, Jonsson algebras in successor cardinals, Israel J. Math., 30 (1978), 57-64.

[Sh 3] S. Shelah, The singular cardinals problem independence results and other problems, A.MS. Abstracts.

[Si] J. Silver, On the singular cardinals problem, Proc. of the International congress of Mathematicians, Vancouver, 1974, Vol.1, 265-268.

TREES, NORMS, AND SCALES

David Guaspari

§0. Introduction

A. In the past few years descriptive set theory has changed from a miscellany to a subject. The main cause has been the study of various determinacy hypotheses - which seems to require a unified, axiomatic approach. This article describes three unifying principles, sets out their relations, and derives from them the best known elementary facts of classical descriptive set theory. (The approach is due primarily to Moschovakis and not all to the author of this paper.)

Sections 1 - 3 consist essentially of one hour talks on each of the notions: trees, norms, scales - the leisure of print allowing the inclusion of a few digressions and, more important, a slightly more abstract approach than is practical in a talk.

Paragraph B of this section fixes notation and some basic terminology. Paragraph C needn't be read immediately. It contains a definition or two that will be referred to later, but consists mainly of a chatty example which, though not necessary to the sequel, may help orient the beginner.

B. $\mathbb{R}$, the set of reals, is ω^ω - the set of functions from ω to ω. We're interested in the descriptive set theory of the <u>pointspaces</u> $\mathbb{R}^m \times \omega^n$; and will use $\mathcal{X}$, $\mathcal{Y}$, $\mathcal{Z}$ for pointspaces and x, y, z, ... for their elements. For the elements of $\mathbb{R}$ we'll also use α, β, γ (ordinals will be denoted by Greek letters from the end of the alphabet); and for elements of ω: i,j,k,m,n. Topologically we think of ω as a discrete space and $\mathbb{R}$ as the topological product of countably many copies of ω.

To avoid any spurious pretence that this article is completely self-contained we'll immediately declare: the reader must at least know the definition of Π^1_n, Σ^1_n, etc. and have a nodding acquaintance with the elementary facts of recursion theory.

From the points of view both of topology and recursion theory there are, really, only two pointspaces: $\mathbb{R}$ and ω. For if $m > 0$, then $\mathbb{R}^m \times \omega^n$ is homeomorphic to $\mathbb{R}$ via a recursive function; and, of course there are also recursive bijections from ω^n to ω. The spaces homeomorphic to $\mathbb{R}$ are of type 1, those homeomorphic to ω of type 0.

A pointset (denoted by A,B,C) is a subset of some pointspace, and a pointclass (denoted by Γ, Γ') a collection of pointsets. The pointclasses of interest will contain subsets of every pointspace. Examples: the (pointclass of) closed sets; the recursive sets; the Π^1_6 sets. The examples to keep in mind, for our purposes, are Π^1_n, Σ^1_n, and Δ^1_n. We will devote attention primarily to subsets of $\mathbb{R}$.

To be at all interesting a pointclass must possess some closure properties. A notational convention: if Op is a logical operation, then Op Γ is the result of applying Op to all pointsets (or combinations of pointsets) in Γ. For example, the sets in $\vee\Gamma$ are obtained as follows: If $A \subseteq \mathcal{X}\times\mathcal{Y}$ and $B \subseteq \mathcal{Y}\times Z$ are in Γ then $C = \{(x,y,z) \mid (x,y) \in A \vee (y,z) \in B\}$ is in $\vee\Gamma$.

We'll use $Q^{<\omega}$, $\forall^\omega$, $\forall^{\mathbb{R}}$ to represent, respectively: bounded numerical quantification (both existential and universal), universal quantification over ω, and universal quantification over $\mathbb{R}$. (Final example: if $A \in \Gamma$, then $B = \{(m,x) \mid \forall n \leq m\, (n,m,x) \in A\}$ is in $Q^{<\omega}\Gamma$.) Γ is closed under Op if Op $\Gamma \subseteq \Gamma$.

Here are some closure properties of Γ if Γ is one of Σ^1_n, Π^1_n, Δ^1_n: (The grouping is for future reference.)

I a) closed under $\wedge$, $\vee$, $Q^{<\omega}$

b) closed under recursive substitution;

i.e.

if $f\colon \mathcal{X} \to \mathcal{Y}$ is total recursive and $A \subseteq \mathcal{Y}$ is in Γ, then $\{x \mid f(x) \in A\}$ is in Γ.

II a) closed under $\forall^\omega$, $\exists^\omega$

b) closed under $\forall^{\mathbb{R}}$ if Γ is Π^1_n under $\exists^{\mathbb{R}}$ if $\sum^1_n$; under neither if Δ^1_n.

<u>Definition</u>

Γ is adequate iff Γ contains all recursive sets and is closed under the operations in I.

Not only are the Π^1_n, etc., adequate, but so also are: $\underset{\sim}{\Pi}^1_n$, $\underset{\sim}{\Sigma}^1_n, \ldots$; Π^o_n, $\sum^o_n, \ldots$; the collection of pointsets recursive in some higher type object ; etc.

One last piece of notation. $\underset{\sim}{\Gamma}$ is the relativization of Γ. Its elements are those sets of form $\{x|(\alpha,x) \in A\}$ for $\alpha \in \mathbb{R}$ and $A \subset \Gamma$. If Γ is adequate, so is $\underset{\sim}{\Gamma}$; and we get the same collection of relativized sets from Γ whether we fix several co-ordinates or just one.

C. Introductory examples (having little to do with the sequel).

The theorems of the sequel have the following look: If Γ has certain closure properties (e.g., is adequate) and certain structure properties (examples follow), then.... (something follows).

Here is an example of a structure property: the parametrization property. Notation : If $B \subseteq \mathcal{X}\times\mathcal{Y}$, then for any $x \in \mathcal{X}$, B_x is $\{y|(x,y) \in B\}$.

<u>Definition</u>

B is $\mathcal{X}$-universal for Γ subsets of $\mathcal{Y}$

iff

1) $B \in \Gamma$ and $B \subseteq \mathcal{X}\times\mathcal{Y}$

2) $\{A \in \Gamma | A \subseteq \mathcal{Y}\} = \{B_x | x \in \mathcal{X}\}$

If B is universal and $A = B_x$ we call an index for A (although it should properly be called a B index for A).

<u>Definition</u>

Γ is $\mathcal{X}$-parametrized iff for every $\mathcal{Y}$ there exists a set which is $\mathcal{X}$-universal for Γ subsets of $\mathcal{Y}$; and, simply, parametrized, if $\mathcal{X}$-parametrized for some $\mathcal{X}$.

One of the elementary facts of descriptive set theory (which will not be proved here) is that:

Each of Π^0_n, Σ^0_n, Π^1_n, Σ^1_n (but not Δ^1_n) is ω-parametrized.

Each of $\underset{\sim}{\Pi}^0_n$, $\underset{\sim}{\Sigma}^0_n$, $\underset{\sim}{\Pi}^1_n$, $\underset{\sim}{\Sigma}^1_n$ (but not $\underset{\sim}{\Delta}^1_n$) is $\mathbb{R}$-parametrized.

In the abstract, we can tidy up parametrizations as follows:

If Γ is adequate and $\mathcal{X}$-parametrized, then Γ is $\mathbb{R}$-parametrized if type $(\mathcal{X}) = 1$, ω-parametrized if type $(\mathcal{X}) = 0$.

Proof. Say type $(\mathcal{X}) = 1$, and that $B \subseteq \mathcal{X} \times \mathcal{Y}$ is an $\mathcal{X}$-parametrization of the Γ subsets of $\mathcal{Y}$. Let $f\colon \mathbb{R} \overset{1-1}{\longleftrightarrow} \mathcal{X}$ be recursive and put $B' = \{(\alpha,y) \mid (f(\alpha),y) \in B\}$. Since the map $(\alpha,y) \longmapsto (f(\alpha),y)$ is recursive and Γ is adequate, $B' \in \Gamma$. And clearly $\{B'_\alpha \mid \alpha \in \mathbb{R}\} = \{B_x \mid x \in \mathcal{X}\}$.

Similarly easy to prove is the fact that:

Γ adequate and parametrized $\Longrightarrow$ so is $\underset{\sim}{\Gamma}$.

The usual parametrizations of, say, the Π^1_n sets, have important uniformity properties (the s.m.n theorems). For example, there is a recursive $f\colon \omega^2 \to \omega$ such that if n is a Π^1 index for A and m a Π^1_n index for B, then $f(m,n)$ is a Π^1_n index for $A \cap B$.

The definition of "parametrized" does not require that the various universal sets be related in any such nice way, but:

If Γ is adequate and parametrized, then there is a collection of universal sets $\{B_\mathcal{Y} \mid \mathcal{Y}$ a pointspace$\}$ having all the uniformity properties one could hope for.

For simplicity let's look at spaces of type 1. For each $\mathcal{Y}$ of type 1 let $f_\mathcal{Y}\colon \mathcal{Y} \overset{1-1}{\longleftrightarrow} \mathbb{R}$ be the "obvious" recursive map. Let $B_\mathbb{R}$ be $\mathcal{X}$ universal for the Γ subsets of $\mathbb{R}$ and obtain all the others by: $B_\mathcal{Y} = \{(x,y) \mid (x,f_\mathcal{Y}(y)) \in B_\mathbb{R}\}$. Since the $f_\mathcal{Y}$'s are nicely related so are the $B_\mathcal{Y}$'s.

Dealing with adequate parametrized pointclasses, then, we are on familiar ground. We can use all the indexing tricks familiar from recursion theory. One example,

Theorem (Hierarchy theorem)

If Γ is adequate and parametrized, then $\neg\Gamma \not\subseteq \Gamma$.

Proof.

The usual one. Let Γ be $\mathcal{X}$-parametrized. Let B be $\mathcal{X}$ universal for Γ subsets of $\mathcal{X}$ and $A = \{x \mid (x,x) \notin B\}$. The adequacy of Γ guarantees the adequacy of $\neg\Gamma$, which in turn guarantees that $A \in \neg\Gamma$. As usual, $A \notin \Gamma$. For if $A \in \Gamma$ and y is an index for A,

$$(y,y) \notin B \iff y \in A \iff y \in B_y \iff (y,y) \in B$$

The reader may simply shrug at what has after all been nothing more than an axiomatization of the well known details of a well known proof. Indeed, the example was chosen for practice, to go down easily. But is has the virtue of focusing the attention in the right places and of avoiding a wild and forbidding notation. And it helps single out the notions to be discussed in the next three sections.

§1. Trees

A. It will be convenient to define "tree on ξ", "tree on $\xi\times\eta$", etc., rather than to provide a single definition "tree on the set X."

For any X, FS(X) is the set of finite sequences from X. We'll use s,t,... variously sub or super scripted for finite sequences. If f is a finite or infinite sequence and $n \in \omega$, then $\bar{f}(n)$ is the finite sequence $f \upharpoonright n$ --i.e., $(f(0),\ldots,f(n-1))$. Notice that $\bar{f}(0)$ is the empty sequence and that n > every element of domain(f) implies that $\bar{f}(n) = f$.

T is a *tree on* ξ iff 1) $\emptyset \neq T \subseteq FS(\xi)$

2) for all $n \in \omega$ and all s, $s \in T \Longrightarrow \bar{s}(n) \in T$.

A *path through* T is a function $f\colon \omega \to \xi$ such that for all n, $\bar{f}(n) \in T$. Finally, [T] is the set of all paths through T.

Trees grow downward and a path through T corresponds to an infinite descending chain in T. (Precisely: partially order T by $s \preceq t$ iff $t \subseteq s$.)

Since a path through a tree on ξ is an element of ${}^{\omega}\xi$, a path through a tree J on $\xi\times\eta$ ought to be a pair (f,g) with $f \in {}^{\omega}\xi$, $g \in {}^{\omega}\eta$, and f and g running through J side-by-side:

J is a *tree on* $\xi\times\eta$ iff 1a) $\emptyset \neq T \subseteq FS(\xi)\times FS(\eta)$

1b) $(s,t) \in T \Longrightarrow$ length of s equals length of t

2) for all $n \in \omega$ and all s and t, $(s,t) \in T \Longrightarrow (\bar{s}(n),\bar{t}(n)) \in T$.

A path through J is a pair $(f,g) \in {}^{\omega}\xi\times{}^{\omega}\eta$ such that for all $n \in \omega$, $(\bar{f}(n),\bar{g}(n)) \in T$. Again, [J] is the set of all paths through J (and $(s,t) \preceq (s',t')$ iff $s \le s'$ and $t \le t'$).

Etc.

It is easy to check:

Lemma 1.1.

$A \subseteq \mathbb{R}^n$ is closed iff $A = [T]$ for some T on ω^n

$A \subseteq \mathbb{R}^n$ is Π^0_1 iff $A = [T]$ for some recursive tree T on ω^n.

B. Let A be a Σ^1_1 subset of $\mathbb{R}$. Then A is the projection of a Π^0_1 subset B of $\mathbb{R}^2$ (and conversely). For any real α let B_α be the α-section of B: $B_\alpha = \{\beta \mid (\alpha,\beta) \in B\}$. So

$$\alpha \in A \iff \exists\beta(\alpha,\beta) \in B$$
$$\iff B_\alpha \neq \emptyset.$$

We can rephrase the above in terms of trees. Let T be a recursive tree on $\omega\times\omega$ such that $B = [T]$. Then for any α the closed set B_α is $[T(\alpha)]$, where $T(\alpha)$ is the following tree on ω: $\{s \mid \exists n(\bar{\alpha}(n),s) \in T\}$. So,

$$\alpha \in A \iff \exists\beta(\alpha,\beta) \in [T]$$
$$\iff [T(\alpha)] \neq \emptyset$$
$$\iff T(\alpha) \text{ is not wellfounded (under } \precsim)$$

Abstracting from this:

Definition 1.2

If T is a tree on $\omega\times\xi$
then $p[T] = \{\alpha \mid T(\alpha) \text{ is not wellfounded}\}$
where $T(\alpha) = \{s \mid \exists n\ (\bar{\alpha}(n),s) \in T\}$.

Since wellfoundedness is absolute (for transitive models of enough set theory) we immediately have:

Theorem 1.3.

"$\alpha \in p[T]$" is an absolute predicate of α,T.

Corollary 1.4.

Σ^1_1 predicates are absolute for transitive models of (enough) set theory.

(To be a little more explicit about the corollary: Given an index for a Σ^1_1 set there is an absolute - in fact, recursive - way to obtain a suitable tree T,... etc.)

Notes:

Here κ is any ordinal, though often only its cardinality matters. Say that A is κ-Souslin if $A = p[T]$ for some tree T on κ. We've shown that ω-Souslin is equivalent to Σ^1_1. Trivially, if κ is the cardinality of $\mathbb{R}$, then every set of reals is κ-Souslin; and we'll indicate a proof (later) that in ZF one can prove outright that all Σ^1_2 sets are $\aleph_1$-Souslin.

C. Perfect set theorems. $A \subseteq \mathbb{R}$ is perfect if it is non-empty and closed, and every point of A is a limit point of A. T, a tree on ω, is a perfect tree if every $s \in T$ splits in T - i.e., has at least two incompatible (with respect to $\leq$) extensions in T. It is easy to see that A is perfect if and only if $A = [T]$ for some perfect tree T.

A perfect set contains a topological copy of ${}^{\omega}2$, so has the cardinality of the continuum. One of the early programs of descriptive set theory was to prove pieces of CH by showing for larger classes Γ that Γ had the perfect set property:

$PSP(\Gamma) =_{df}$ every uncountable element of Γ contains a perfect subset. (There exist sharper, more effective, versions of the PSP. See below.)

We'll show below that ZF proves $PSP(\underset{\sim}{\Sigma}^1_1)$ - a classical result. Godel observed that V = L implies $\neg PSP(\Pi^1_1)$. A theorem due independently to Mansfield and to Solovay says that $PSP(\underset{\sim}{\Pi}^1_1)$ is equivalent to the statement: for every α only countably many reals are constructible from α.

Let's briefly consider the classical proof of PSP(closed sets). Start with $A = [T]$ and try to find a perfect tree $T^* \subseteq T$ - for then the perfect set $[T^*]$ will be contained in A. We attempt to obtain T^* by pruning T of its non-splitting nodes. For any tree J on ω let J' be $\{s \mid s \text{ splits in } J\}$. (Notice that J is perfect iff J' = J.) One such pruning may not suffice, for it's possible that some node which splits in J no longer splits in J'. Put $T^0 = T$; $T^{\xi+1} = (T^{\xi})'$; and for limit λ, $T_{\lambda} = \bigcap_{\xi<\lambda} T^{\xi}$. For some $\eta < \aleph_1$ this stops: $T^{\eta} = T^{\eta+1}$ =, say, T^*. If $T^* = \emptyset$, then, presumably, we can show that [T] was countable to begin with. Proof: We've reduced [T] to $\emptyset$ in countably many steps. All we need to know is that at each step only countably many paths were discarded (i.e., $[T^{\xi}]\backslash[T^{\xi+1}]$ is countable). But all the paths in $[T^{\xi}]\backslash[T^{\xi+1}]$ are isolated points of $[T^{\xi}]$.

Here is an ultramodern version of that argument:

Theorem 1.5.

Let M be a transitive model of enough set theory and T, a tree on $\omega\times\xi$, be an element of M. Then,

$p[T] \not\subseteq M \Longrightarrow p[T]$ contains a perfect set.

Before proving 1.5 we will use it to deduce $PSP(\underset{\sim}{\Sigma}^1_1)$ (in a rather heavy-handed way) and the Mansfield-Solovay theorem:

Corollary 1.6.

$PSP(\underset{\sim}{\Sigma}^1_1)$

Proof

Suppose A is a $\sum_1^1$ set and T is a tree $\omega\times\omega$ with $A = p[T]$. Let M be a countable transitive model of set theory with $T \in M$. Then if A is uncountable $p[T] \underset{\sim}{\not\subset} M$ - and so A contains a perfect subset.

Corollary 1.7 (Mansfield; Solovay)

Any $\sum_2^1$ set containing a non-constructible real contains a perfect set.

"proof"

Show that any $\sum_2^1$ set is $p[T]$ for some $T \in L$. (This is non-trivial. The tree in question is usually called the Schoenfield tree of the $\sum_2^1$ set.)

Note:

From 1.7 it follows fairly easily that PSP$(\underset{\sim}{\Pi}_1^1)$ <==> PSP$(\underset{\sim}{\Sigma}_2^1)$ <==> for any real α, $L[\alpha] \cap \mathbb{R}$ is countable.

Proof of 1.5

For any tree J on $\omega\times\xi$ say that (s,t) splits in J if there exist (s_0,t_0) and (s_1,t_1) in J such that $(s_i,t_i) \precsim (s,t)$ and s_0 and s_1 are incompatible; and let $J' = \{(s,t) | (s,t)$ splits in $J\}$.

Suppose now that $T \in M$ is a tree on $\omega\times\xi$ and let $T^0 = T$; $T^{\nu+1} = (T^\nu)'$; and, for limit λ, $T^\lambda = \bigcap_{\nu<\lambda} T^\nu$. Since the operation $J \mapsto J'$ is absolute the sequence of T^ξ's is definable in M (through the ordinals of M). Let η be the least ordinal satisfying $T^\eta = T^{\eta+1}$. Then $\eta \in M$. (Proof: M must satisfy "the sequence of T^ν's is eventually constant"; so for some $\eta^* \in M$, M satisfies "$T^{\eta^*} = T^{\eta^*+1}$". But then it must be true that $T^{\eta^*} = T^{\eta^*+1}$. So $\eta \le \eta^*$.) Consider two cases:

Case 1. $T^\eta \ne \emptyset$.

It's easy to check that $p[T^\eta]$ is a subset of $p[T]$ containing a perfect set. (There is no reason to believe that $p[T^\eta]$ is itself closed.)

Case 2. $T^{\eta} = \emptyset$

We'll show $p[T] \subseteq M$. Suppose $\alpha \in p[T]$, and choose any f such that $(\alpha,f) \in [T]$. Since (α,f) is an element of $[T^{o}]$ but not of $[T^{\eta}]$, there must be some ν satisfying: $(\alpha,f) \in [T^{\nu}]\setminus[T^{\nu+1}]$. Since $(\alpha,f) \notin [T^{\nu+1}]$ there is some n for which $(s,t) =_{df} (\bar{\alpha}(n),\bar{f}(n)) \in T^{\nu+1}$, which means that (s,t) does not split in T^{ν}. Then: (1) If (s,t) and (s_1,t_1) are elements of T^{ν} extending (s,t), s_o and s_1 must be compatible. (2) Every initial segment s' of α^{o} satisfies: there exists t' such that $(s',t') \in T^{\nu}$ and $(s',t') \precsim (s,t)$. So $\alpha = \cup\{s' \mid \exists t'[(s',t') \in T^{\nu}$ and $(s',t') \precsim (s,t)]\}$. But this definition is absolute and therefore $\alpha \in M$.

A closer look at the argument gives:

Theorem 1.8

The "M" of 1.5 can be taken to be the least admissible set containing T.

(Note: The proof of 1.5 presented above is a little too crude. If M is merely T-admissible it is not, it turns out, guaranteed that $\eta \in M$, but only that $\eta \leq On \cap M$. That, however, is good enough to carry through the rest of the argument.)

Corollary 1.9 (Harrison) (A "lightface" perfect set theorem)

If a Σ^1_1 set contains a non-hyperarithmetic real, then it contains a perfect subset.

Proof

Let T be an infinite recursive tree on $\omega\times\omega$ and apply 1.8 in light of the following facts: The least T-admissible set is $L_{\omega_1^{cK}}$. The reals in $L_{\omega_1^{cK}}$ are precisely the hyperarithmetic reals.

(Note: The proof presented for 1.9, like that for 1.6, is much more high powered than necessary.)

D. A Miscellany of other applications.

The Kunen-Martin theorem: If $A \subseteq \mathbb{R}^2$ is a wellfounded relation and a κ-Souslin set, then the ordinal length of A is less than κ^+.

One corollary of this is that every $\undertilde{\Sigma}^1_2$ wellordering has length less than $\aleph_2$. So, if there exists a $\undertilde{\Sigma}^1_2$ wellordering of $\mathbb{R}$, then CH holds. (Mansfield has proved the much stronger result that if $\mathbb{R}$ admits a Σ^1_2 wellordering, then all reals are constructible. On the other side, Harrington has shown that it is consistent with ZFC that the reals have a Δ^1_3 wellordering but that CH be false.)

Generalizing the Souslin-Kleene Theorems: A is κ-Borel if A is in the least class containing all clopen sets and closed under unions and intersections of fewer than κ sets. (So "Borel" is "$\aleph_1$-Borel".) The classical results (and proofs) that

i) Borel $\subseteq \undertilde{\Sigma}^1_1$

ii) disjoint $\undertilde{\Sigma}^1_1$ sets can be separated by Borel sets;

i.e.

i) ω^+-Borel $\subseteq$ ω-Souslin

ii) disjoint ω-Souslin sets can be separated by ω^+-Borel sets;

become

i') κ^+-Borel $\subseteq$ κ-Souslin

ii') disjoint κ-Souslin sets can be separated by κ^+-Borel sets.

Since ω^+-Borel and ω-Souslin are equivalent to $\undertilde{\Delta}^1_1$ and $\undertilde{\Sigma}^1_1$ it is natural to look for other correlations between the hierarchy of Souslin operations and the analytical hierarchy. Nontrivial results have been obtained from AD.

§2. Norms

A. From the representation theorem for Σ^1_1 sets we know that for any Π^1_1 set A there is a recursive tree T such that

$x \in A \iff T(x)$ is wellfounded.

I assume that the reader has, at some time in his life, seen the succession of coding tricks which transforms this into the basic representation theorem:

Theorem 2.1

A is Π^1_1 iff for some total recursive function f into $\mathbb{R}$,

$$x \in A \iff f(x) \text{ codes a wellordering.}$$

By "α codes a relation B on ω" we mean that $B = \{(m,n) \mid \alpha(\langle m,n\rangle) = 0\}$, where $\langle , \rangle : \omega^2 \longleftrightarrow \omega$ is some fixed recursive pairing function. For future reference, the relation coded by β will be denoted by $\leq_\beta$; WO $= \{\beta \mid \beta$ codes a wellordering$\}$; and for $\beta \in$ WO, $|\beta|$ = the ordinal coded by β = the order type of $\leq_\beta$.

The basic representation theorem provides us with a natural way of attaching ordinals to the elements of a Π^1_1 set. In the setting of Theorem 2.1, define σ: A $\longrightarrow$ On by $\sigma(x) = |f(x)|$. Such a map induces a relation $\le_\sigma$ on A --<u>a prewellordering</u> --by

(*) $\alpha \le_\sigma \beta \iff \sigma(\alpha) \le \sigma(\beta)$.

<u>Definition 2.2</u>

A map from A into On is a <u>norm</u> on A. A relation $\le_\sigma$ obtained from a norm σ as in (*) is a <u>prewellordering of A</u>. (Alternatively: a prewellordering satisfies all the requirements for a well-ordering except, possibly, anti-symmetry.)

Of course every set admits trivial norms. The interest lies in those norms whose associated prewellorderings are simply definable. The most important property of the pwo's provided by the basic representation theorem is this: Not only are they Π^1_1 relations, but also their initial segments are $\underset{\sim}{\Delta}^1_1$ in a uniform way.

<u>Theorem 2.3</u> (pwo theorem for Π^1_1)

Let A be Π^1_1, σ the norm obtained from the basic representation theorem. Then there is a Π^1_1 relation $\le^\Pi$ and a Σ^1_1 relation $\le^\Sigma$ such that:

$$(^*_*)\begin{cases} \text{For all } y \in A \text{ and all } x, \\ \qquad x \in A \text{ and } \sigma(x) \le \sigma(y) \iff x \le^\Sigma y \\ \qquad\qquad\qquad\qquad\qquad \iff x \le^\Pi y. \end{cases}$$

(In future we'll refer to $(^*_*)$ as 'the set-theoretical requirements on $\le^\Sigma$ and $\le^\sigma$'.)

<u>Proof of 2.3</u>

Let f be as in the basic representation theorem. Here's how to define $\le^\Sigma$: $x \le^\Sigma y$ iff $\exists\, \alpha$ (α is an order preserving map from $\le_{f(x)}$ into $\le_{f(y)}$). Notice that if $y \in A$, then $\le_{f(y)}$ is a wellordering and so the existence of such an α guarantees that $\le_{f(x)}$ is a wellordering (hence that $x \in A$) and of length less than or equal to that of $\le_{f(y)}$. If $y \notin A$ we don't care in the least which x's satisfy $x \le^\Sigma y$. The definition is Σ^1_1 because the expression in parentheses involves only recursive operations and number quantifiers. For the record: $x \leq^{\Pi} y$ iff $x \in A$ and $\neg\exists\alpha$ (α embeds $\leq_{f(y)}$ as a proper segment of $\leq_{f(x)}$).

We'll now abstract from this. (Use $\breve{\Gamma}$ to stand for the dual class of Γ - what we've previously called $\neg\Gamma$.)

Definition 2.4

A norm σ on A is a Γ-norm iff there exist relations $\leq^{\Gamma}$ in Γ and $\leq^{\breve{\Gamma}}$ in $\breve{\Gamma}$ such that

for all $y \in A$ and all x,

$$x \in A \text{ and } \sigma(x) \leq \sigma(y) \iff x \leq^{\Gamma} y$$
$$\iff x \leq^{\breve{\Gamma}} y.$$

Definition 2.5

$PWO(\Gamma)$ iff every set in Γ has a Γ norm.

B. Some consequences of PWO

Reduction and separation

$Red(\Gamma)$ says that given any $A_0, A_1 \in \Gamma$ there exist $A_0', A_1' \in \Gamma$ satisfying:

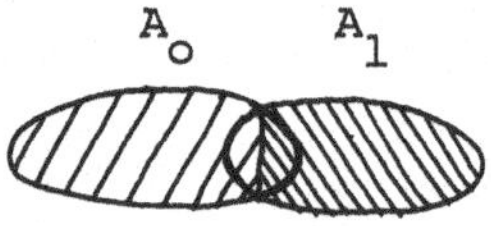

A_o': ///

A_1': \\\\\\\\\\\\;

i.e., $A_i' \subseteq A_i$; $A_0' \cap A_1' = \emptyset$; and $A_0' \cup A_1' = A_0 \cup A_1$.

$Sep(\breve{\Gamma})$ says that given any disjoint $A, B \in \Gamma$ there exists $C \in \Gamma \cap \breve{\Gamma}$ satisfying:

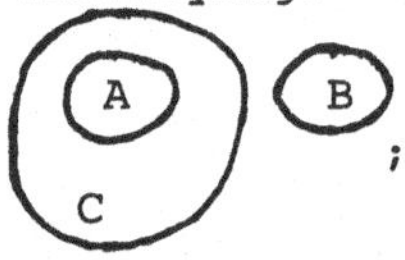

; i.e., $A \subseteq C$ and $B \cap C = \emptyset$.

The following fact is a consequence of propositional logic:

Theorem 2.6

$Red(\Gamma) \implies Sep(\breve{\Gamma})$.

Our first application of PWO is to prove Red and Sep.

Theorem 2.7

Γ adequate and $PWO(\Gamma) \implies Red(\Gamma)$, hence $Sep(\breve{\Gamma})$.

Corollary 2.8

$\text{Red}(\Pi^1_1)$, $\text{Sep}(\Sigma^1_1)$.

Proof of 2.7

Let $A_0, A_1 \in \Gamma$. The problem is: if $x \in A_0 \cap A_1$ do we assign x to A'_0 or to A'_1? We'd like to say: take norms σ_i and put x in A'_0 just in case $\sigma_0(x)$, "the ordinal which puts x in A_0", is less than $\sigma_1(x)$. The A'_i would satisfy the desired set theoretical requirements but needn't be elements of Γ because the pwo property tells us nothing about comparisons between different norms. Apply a simple trick. Let $B = \{(0,x) \mid x \in A_0\} \cup \{(1,x) \mid x \in A_1\}$ - thus B is a Γ subset of $\omega \times \mathbb{R}$ - and let σ be a Γ norm on B. Put

$$x \in A'_0 \iff x \in A_0 \text{ and } (x \in A_1 \implies (0,x) <_\sigma (1,x))$$

$$x \in A'_o \iff x \in A_1 \text{ and } (x \in A_0 \implies (1,x) \leq_\sigma (0,x)).$$

In order to illustrate a characteristic point about pwo's let's show in detail that $A'_o \in \Gamma$.

$$\begin{aligned} x \in A'_0 &\iff x \in A_0 \text{ and } \neg(x \in A_1 \text{ and } (0,x) \not<_\sigma (1,x)) \\ &\iff x \in A_0 \text{ and } \neg(x \in A_1 \text{ and } (1,x) \leq_\sigma (0,x)) \\ &\iff x \in A_0 \text{ and } \neg((1,x) \leq^{\breve{\Gamma}} (0,x)). \end{aligned}$$

The point is that: The statements "$(0,x) \not<_\sigma (1,x)$", "$(1,x) \leq^{\breve{\Gamma}} (0,x)$" are not necessarily equivalent to one another for all possible x. However, if $x \in A_0$, then the statements "$x \in A_1$ and $(0,x) \not<_\sigma (1,x)$", "$x \in A_1$ and $(1,x) \leq_\sigma (0,x)$", and "$(1,x) \leq^{\breve{\Gamma}} (0,x)$" are equivalent. (To show that A'_1 is Γ it is most convenient first to establish the following fact: let $x <_\sigma y$ iff $\sigma(x) < \sigma(y)$. Then the initial segments of $<_\sigma$ are also uniformly elements of $\Gamma \cap \breve{\Gamma}$).

Propogation of PWO

Theorem 2.8

Suppose Γ is adequate and $\forall^{\mathbb{R}} \Gamma \subseteq \Gamma$.

Then, $\text{PWO}(\Gamma) \implies \text{PWO}(\forall^{\mathbb{R}} \Gamma)$.

Proof

Suppose that $A = \{x \mid \exists\beta(x,\beta) \in B\}$ with $B \in \Gamma$. Let σ be a Γ norm on B and define a norm ψ on A as follows:

$$\psi(x) = \inf\{\sigma(x,\beta) \mid (x,\beta) \in B\}.$$

Then it is easy to check that ψ is an $\exists^{\mathbb{R}}\Gamma$ norm on A.

Corollary 2.9

$\mathrm{PWO}(\Sigma^1_2)$, $\mathrm{Red}(\Sigma^1_2)$, $\mathrm{Sep}(\Pi^1_2)$.

It is now tempting to suppose that one could obtain $\mathrm{PWO}(\Pi^1_3)$ by repeating the last trick with "sup" in place of "inf". It is midly instructive to try that and see where it fails. The full story is:

i) If Γ is adequate and parametrized (see §0.C) then $\mathrm{PWO}(\Gamma)$ and $\mathrm{PWO}(\check{\Gamma})$ cannot both hold. In particular, it cannot happen that both Σ^1_n and Π^1_n have PWO.

ii) (Addison) $V = L \Longrightarrow \mathrm{PWO}(\Sigma^1_n)$ for $n \geq 2$.

iii) (Addison-Moschovakis; Martin) Projective determinancy implies that PWO holds for each Π^1_{odd} and Σ^1_{even}.

iv) (Harrington) Any pattern for PWO consistent with $\mathrm{PWO}(\Pi^1_1)$ and theorem 2.8 is possible (in some model of set theory).

Boundedness and the lengths of pwo's

Notation: Abbreviate $\Gamma \cap \check{\Gamma}$ by Δ; $\underset{\sim}{\Gamma}$ is the boldface version of Γ.

Theorem 2.10 (Boundedness theorem for subsets of $\mathbb{R}$)

Suppose that Γ is adequate and $\forall^{\mathbb{R}}\Gamma \subseteq \Gamma$. Let $A \in \Gamma$ be such that $A \in \Gamma \setminus \underset{\sim}{\Delta}$; and let σ be a Γ norm on A.

Then every $\underset{\sim}{\check{\Gamma}}$ subset of A is bounded in σ; i.e., if $B \in A$ and $B \in \underset{\sim}{\check{\Gamma}}$, there is an $\alpha \in A$ such that $\beta \in B \Longrightarrow \sigma(\beta) < \sigma(\alpha)$.

Proof

Suppose B is a counterexample. Then we get an immediate contradiction by providing the following Γ definition of A:

$$\alpha \in A \Longleftrightarrow \exists\beta(\beta \in B \text{ and } \alpha \leq^{\check{\Gamma}} \beta).$$

Notes: There is an analogous theorem for subsets of ω (or any other pointspace). The requirement that $\forall^{\mathbb{R}} \subseteq \Gamma$ cannot merely be omitted, for the boundedness theorem is not true of Σ^1_2.

Corollary 2.11

If B is a $\underset{\sim}{\Sigma}_1^1$ subset of WO, then $\sup\{|\alpha| \mid \alpha \in B\}$ is countable.

Proof

We need to know: (1) that WO is Π_1^1 but not $\underset{\sim}{\Delta}_1^1$, (2) that the map sending α to $|\alpha|$ is a Π_1^1 norm on WO. (2) is proved by consulting the proof of 2.3 (the pwo theorem for Π_1^1). If WO were $\underset{\sim}{\Delta}_1^1$ the basic representation theorem (theorem 2.1) would immediately imply that every $\underset{\sim}{\Pi}_1^1$ set is $\underset{\sim}{\Delta}_1^1$, contradicting the hierarchy theorem.

The boundedness theorem, simple as it is, nonetheless allows us to get a handle on the particular ordinals which appear in norms. To be able to talk sensibly about such things say that the length of a norm is the order type of its range. (So if, as we can without loss of generality do, we require a norm to be onto an ordinal, then its length is its range.) Notice that WO is a complete Π_1^1 set in the sense that every Π_1^1 set is a recursive preimage of WO (and every $\underset{\sim}{\Pi}_1^1$ set a continuous preimage). The natural Π_1^1 norm on WO has length $\aleph_1$ and so all norms on Π_1^1 sets obtained from the basic representation theorem have length $\leq \aleph_1$. In fact, every Π_1^1 norm on WO - or on any complete Π_1^1 set - has length exactly $\aleph_1$; and every Π_1^1 norm has length $\leq \aleph_1$.

We'll state the facts solely for Π_1^1, but in a way that suggests the correct generalizations:

(1) length of any Π_1^1 norm on a complete Π_1^1 subset of $\mathbb{R}$
$= $ longest possible Π_1^1 norm
$= \sup\{\xi \mid \xi$ is the length of a $\underset{\sim}{\Delta}_1^1$ pwo of $\mathbb{R}\}$
$(= \aleph_1)$

(2) length of any Π_1^1 norm on a complete subset of ω
$=$ longest possible Π_1^1 norm on a subset of ω
$= \sup\{\xi \mid \xi$ is the length of a Δ_1^1 pwo of $\omega\}$
$(= \omega_1^{CK})$

D. PWO and determinacy

The definitions of "game" and "determinacy" are assumed known. We'll show how to deduce PWO(Π_3^1) from the axiom of determinacy (in fact, from the assumption that $\underset{\sim}{\Delta}_2^1$ games are determined). Here "3" is a generic odd integer: The proof axiomatizes, and so, in conjunction with 2.8, we obtain the pattern mentioned before: PD implies that PWO holds for Π_{2n+1}^1

and for $\sum^1_{2n+2}$. (This result is due independently to Martin and to Moschovakis and Addison. The proof of Moschovakis and Addison is presented below.)

Suppose that A is $\sum^1_2$ and $B = \{x | \forall\alpha(\alpha,x) \in A\}$. Let σ be a $\sum^1_2$ norm on A. As mentioned before, setting $x \lesssim y$ to mean that $\sup\{\sigma(\alpha,x) | (\alpha,x) \in A\} \leq \sup\{\sigma(\alpha,y) \in A\}$ doesn't yield a Π^1_3 norm on B. The idea of the proof is to effectivize this. We'll put $x \lesssim y$ if given any α, there is a strategy for producing a β with $\sigma(\alpha,x) \leq \sigma(\beta,y)$.

Consider first the games $G_{x,y}$ defined <u>only when</u> $x,y \in B$. Player I plays α, player II plays β, and

II wins iff $(\alpha,x) \leq_\sigma (\beta,y)$.

Now put $x \lesssim y$ iff $x,y \in B$ and II has a winning strategy in $G_{x,y}$. Finally, let $x < y =_{df} x \lesssim y$ and $\neg y \lesssim x$. From now on ASSUME: $\underset{\sim}{\Delta}^1_1$ determinacy.

<u>Sublemma 0.</u>

For any $x,y \in B$, $G_{x,y}$ is determined.

For any $x,y \in B$, $x < y$ iff I has a winning strategy in $G_{y,x}$.

<u>Proof</u>

Let $x,y \in B$. It's trivial to check that the set of winning plays for I is $\underset{\sim}{\Delta}^1_2$. So $G_{x,y}$ is determined. For the second part, note first that: I wins G if II does not win $G_{y,x}$ iff $\neg y \lesssim x$. So we immediately have that $x < y \Longrightarrow$ I wins $G_{y,x}$. Suppose, on the other hand, that I wins $G_{y,x}$. Then $\neg y \lesssim x$ and so to complete the proof that $x < y$ we need only check that $x \lesssim y$ - i.e., that II has a winning strategy in $G_{x,y}$. So let τ be a winning strategy for I in $G_{x,y}$ and consider the strategy for II in $G_{x,y}$ indicated by the following diagram. The wavy lines represent plays made in accordance with strategy τ. The run of the game $G_{y,x}$ indicated on the right half of the page is merely player II's scratch paper while he pursues his strategy for $G_{x,y}$.

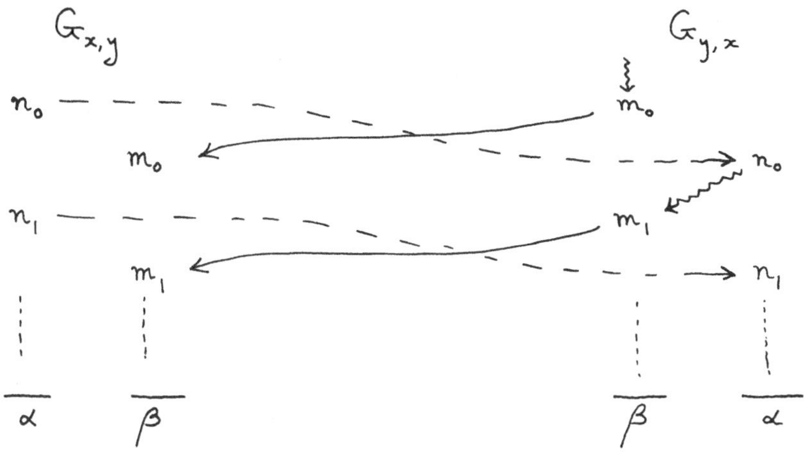

The game goes like this: I plays n_0. II temporarily ignores that play, consults τ, and plays $m_0 = \tau(\emptyset)$, the opening move of τ in the game $G_{y,x}$. I next plays n_1. II now enters onto his scratch pad I's <u>first</u> move. He copies n_0 into the game $G_{y,x}$ - i.e., he presents τ with the sequence of plays $<m_o,n_o>$, then takes τ's response, $m_1 = \tau(<m_o,n_o>)$, and plays it in his own game. Etc. The result is that (α,β) is played in $G_{x,y}$ and (β,α) is a play of $G_{y,x}$ in which the first player has throughout employed strategy τ. So (β,α) is a win for the first player in $G_{y,x}$ - and therefore, because x and y are elements of B, $(\alpha,x) <_\sigma (\beta,y)$; which means that player II has won his play of $G_{x,y}$.

<u>Sublemma 1</u>

$\lesssim$ is a prewellordering

<u>Proof</u>

a) For all $x \in B$, $x \lesssim x$.

proof: A winning strategy for II in $G_{x,x}$ is: Copy the last move of I.

b) $x \lesssim y$ and $y \lesssim z \implies x \lesssim z$

proof: Suppose that II has winning strategies in $G_{x,y}$ and $G_{y,z}$. The diagram indicates a winning strategy for II in $G_{x,z}$.

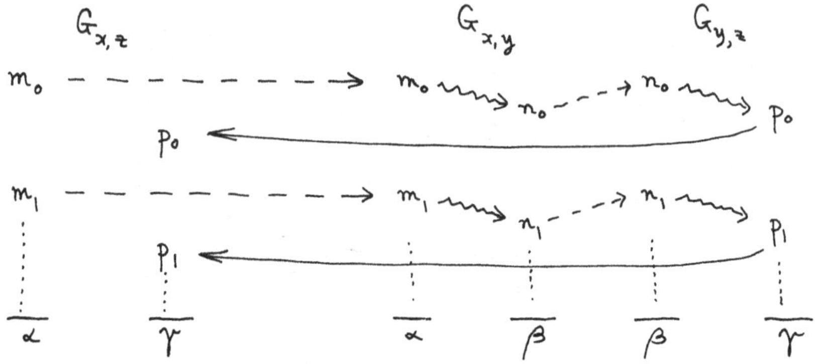

Since (α,β) and (β,γ) are winning plays for II in, respectively $G_{x,y}$ and $G_{y,z}$ we must have $(\alpha,x) \leq_\sigma (\beta,y) \leq_\sigma (\gamma,z)$. Thus II has won the play of $G_{x,z}$.

c) For all $x,y \in B$, $x \lesssim y$ or $y \lesssim x$.

Proof: Each of the following sentences implies the next. $\neg x \lesssim y$. II does not win $G_{x,y}$. I wins $G_{y,x}$. $y < x$. $y \lesssim x$.

d) $\lesssim$ is wellfounded.

Proof: Suppose that $x_0 > x_1 > x_2 > \ldots$ with each $x_i \in B$. Then I must win each of the games $G_{x_i,x_{i+1}}$. (Notice that we're not using sublemma 0.) So consider the plays of these games indicated by the diagram.

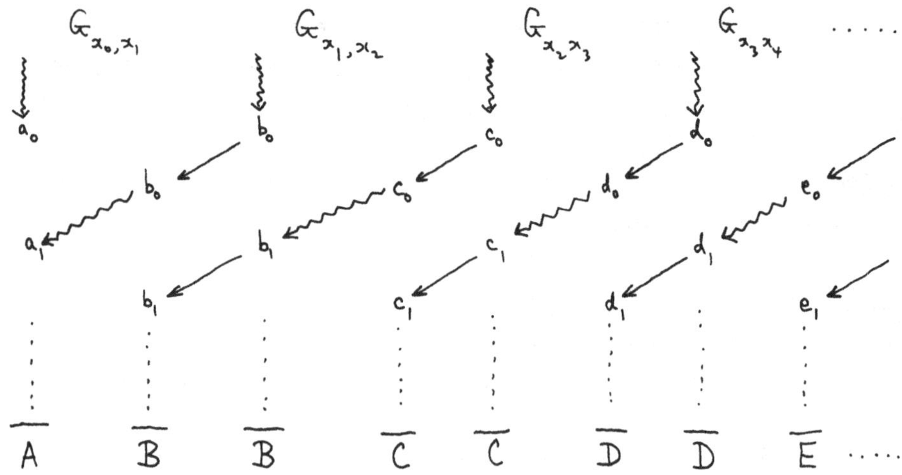

Since I wins each of these games we must have $\sigma(x_0,A) > \sigma(x_1,B)$ from G_{x_0,x_1}; z $\sigma(x_1,B) > \sigma(x_2,C)$ - from G_{x_1,x_2}; etc. Therefore $\leq_\sigma$ is not wellfounded, which is a contradiction.

Sublemma 2

$\precsim$ comes from a Π^1_3 norm

Proof

Let θ be a norm associated with $\precsim$. First notice that $\precsim$ is, outright, a Π^1_3 set - hence we can take $\leq_\theta^{\Pi^1_3}$ to be $\precsim$. For,

$x \precsim y \iff x,y \in B$ and II has a winning strategy in $G_{x,y}$

$\iff x,y \in B$ and I has no winning strategy in $G_{x,y}$

$\iff x,y \in B$ and $\forall\tau\exists\beta\ ((\tau*\beta,x) \leq_\sigma (\beta,y))$

where $\tau*\beta$ is the sequence of moves of player I which results when he uses the strategy τ and II plays β. (To be exact we should say that "τ" varies not over strategies but over reals which encode strategies in some effective way.)

We can hardly expect that $\precsim$ also be Σ^1_3. To obtain the "Σ-side" of $\precsim$, consider the games $H_{x,y}$ defined for all x and y by:

I plays α, II plays β, and II wins iff 1) $(\beta,y) \notin A$

or

2) $(\alpha,x),(\beta,y) \in A$

and $(\alpha,x) \leq_\sigma (\beta,y)$.

The following facts are easy to check: (a) If $y \notin B$, then II has a win in $H_{x,y}$. (b) If $y \in B$ and $x \notin B$, then I has a win in $H_{x,y}$. (c) If $x,y \in B$ then $H_{x,y}$ is the same as $G_{x,y}$. These guarantee first that each $H_{x,y}$ is determined, and second that $H =_{df} \{(x,y) \mid \text{II has a winning strategy in } H_{x,y}\}$ satisfies the set theoretical requirements for $\leq_\theta^{\Sigma^1_3}$.

The proof of the sublemma is concluded by making the routine check that H is in fact Σ^1_3.

What this argument has shown is:

Theorem 2.12 (Martin; Addison-Moschovakis)

Assume determinacy for $\underset{\sim}{\Delta}$ sets and that Γ is adequate. Then,

$$\exists^{\mathbb{R}}\Gamma \subseteq \Gamma \text{ and } \mathrm{PWO}(\Gamma) \Longrightarrow \mathrm{PWO}(\forall^{\mathbb{R}}\Gamma).$$

§3. Scales

A. We're going to glue together the notions of §§1 and 2 to produce the stronger notion of a scale. The idea came from analyzing the argument behind the classical proof of the Π^1_1 uniformization theorem (which says that every Π^1_1 subset of $\mathbb{R}^2$ has a Π^1_1 selection function), and is due to Moschovakis. We'll begin by considering a simple selection problem: If T is a tree on some ordinal ξ, how can we select in some canonical way an element of [T]? One answer: Choose its leftmost branch, h^T. I.e.

$$h^T(0) = \text{least } \eta < \xi \text{ such that for some } f \in [T],\ f(0) = \eta$$

$$h^T(n+1) = \text{least } \eta < \xi \text{ such that for some } f \in [T],\ \langle h^T(0),\ldots,h^T(n),\eta\rangle \subseteq f.$$

Since $\bar{h}^T(n) \in T$ for every n, we must have $h^T \in [T]$. Further, h^T is leftmost in the sense that it is the least element of T with respect to the lexicographical order of ${}^{\omega}\xi$:

$$f <_{lex} g \text{ iff } f \neq g \text{ and, letting } n \text{ be the first input at which they differ, } f(n) < g(n).$$

Although $<_{lex}$ is not a wellordering of ${}^{\omega}\xi$ (nor, usually, of [T]) nonetheless h^T is easily seen to be the least element of [T].

Notice also that the definition of h^T is absolute: For any s let $T[s] = \{t \in T \mid t \text{ is compatible with } s\}$. Then $h^T(n)$ = the least η such that $T[\bar{h}^T(n)^\frown\eta]$ is not wellfounded.

If T is a tree on $\omega\times\xi$ we can easily extend the same trick to $A = p[T]$: For any (α,f) let $\alpha\diagup f$ read "α shuffle f" be the sequence $(f(0), \alpha(0), f(1), \alpha(1),\ldots)$. There are technical reasons for beginning with $f(0)$ rather than with $\alpha(0)$. We choose an element of p[T] by first choosing $(\alpha^*,f^*) \in [T]$ so that $\alpha^*\diagup f^*$ is lexicographically least, then dealing out α^*. When examining α we needn't survey all (α,f) for all possible f, but only those pairs of form (α, h^T_α), where h^T_α is the leftmost branch of $T(\alpha)$.

We want to define α^* "internally" - as we previously showed how to generate the paths h^T. This is straightforward, but it is in fact convenient to be slightly roundabout. By means of the h^T_α's we're provided with a sequence of norms $\vec{\sigma} = (\sigma_0,\sigma_1,\ldots)$ on A as follows:

(**) $\sigma_i(\alpha) = h^T_\alpha(i)$

We now define:

$$(+)\quad \begin{cases} \xi_0 = \text{least ordinal in range } (\sigma_0) \\ \kappa_0 = \text{least } n \in \omega \text{ such that for some } \alpha \in p[T], \\ \qquad (\sigma_0(\alpha) = \xi_0 \text{ and } \alpha(0) = n); \\ \text{and inductively,} \\ A_{m+1} = \{\alpha \mid (\alpha, \vec{\sigma}(\alpha)) \precsim (\langle k_0,\ldots,k_m\rangle, \langle \xi_0,\ldots,\xi_m\rangle)\} \\ \xi_{m+1} = \inf\{\sigma_{m+1}(\alpha) \mid \alpha \in A_{m+1}\} \\ k_{m+1} = \inf\{\alpha(m+1) \mid \alpha \in A_{m+1} \text{ and } \sigma_{m+1}(\alpha) = \xi_{m+1}\}. \end{cases}$$

Here, $\vec{\sigma}(\alpha)$ is the sequence $(\sigma_0(\alpha), \sigma_1(\alpha), \ldots)$.

Now it is trivial to check that $\alpha^* = (k_0, k_1, \ldots)$ and $h^T_{\alpha^*} = f^* = (\xi_0, \xi_1, \ldots)$.

This sequence of norms also has an interesting limit property. To state this it will help to introduce one piece of terminology. Say that a sequence $\langle \alpha_i \mid i<\omega\rangle$ <u>converges in norm to</u> $\vec{\lambda} = (\lambda_0, \lambda_1, \ldots)$ if for each n the sequence $\langle \sigma_n(\alpha_i) \mid i<\omega\rangle$ becomes eventually constant with value λ_n.

<u>Limit Property</u>: If $\langle \alpha_i \mid i<\omega\rangle$ is a sequence of reals in A which converges to α (in the ordinary sense) and converges in norm to $\vec{\lambda}$, then

1) $\alpha \in A$

2) $\vec{\sigma}(\alpha) \leq_{lex} \vec{\lambda}$.

(We'll call conclusion (1) the <u>closure</u> property and conclusion (2) the <u>semi-continuity</u> property.)

<u>Proposition 3.0</u>

The sequence $\vec{\sigma}$ defined in (**) has the limit property.

<u>Proof</u>

Suppose that $\langle \alpha_i \mid i<\omega\rangle$ converges to α and converges in norm to $\vec{\lambda}$.

We'll show $\alpha \in A$ by showing that $(\alpha, \vec{\lambda}) \in [T]$. So, letting $n \in \omega$, we want $(\bar{\alpha}(n), \langle \lambda_0, \ldots, \lambda_{n-1}\rangle) \in T$. For some large enough i, $\bar{\alpha}_i(n) = \vec{\alpha}(n)$ and $\langle \sigma_0(\alpha_i), \ldots, \sigma_{n-1}(\alpha_i)\rangle = \langle \lambda_0, \ldots, \lambda_{n-1}\rangle$. Since $(\alpha_i, \vec{\sigma}(\alpha_i)) = (\alpha_i, h^T_{\alpha_i})$ is a path through T, $(\bar{\alpha}_i(n), \langle \sigma_0(\alpha_i), \ldots, \sigma_{n-1}(\alpha_i)\rangle) \in T$ and we're done. That

$\vec{\sigma}(\alpha) \leq_{lex} \vec{\lambda}$ is immediate because $\vec{\sigma}(\alpha)$ is (by definition) the leftmost branch of $T(\alpha)$.

Definition 3.1a

A scale on A is a sequence of norms on A which has the limit property.

(Note: in some discussions a scale is supposed to satisfy the stronger requirement that "$\vec{\sigma}(\alpha)$ be pointwise less than or equal to $\vec{\lambda}$".)

Suppose, conversely, that we start with a scale $\vec{\sigma}$ on A. We can then associate with that scale a tree T such that $A = p[T]$.

Definition 3.1b

If $\vec{\sigma}$ is a scale on A, the tree associated with $\vec{\sigma}$ is

$\{(\bar{\alpha}(n), \langle\sigma_0(x),\ldots,\sigma_{n-1}(x)\rangle) \mid n \in \omega \text{ and } \alpha \in A\}$.

Lemma 3.2

If $\vec{\sigma}$ is a scale on A and T its associated tree, then $A = p[T]$.

Proof

If $\alpha \in A$, then by the definition of T each proper initial segment of $(\alpha,\vec{\sigma}(\alpha))$ is in T, so $\vec{\sigma}(\alpha)$ is a path through $T(\alpha)$ and $\alpha \in p[T]$.

Suppose now that $\alpha \in p[T]$, and choose some path f through $T(\alpha)$. By the definition of T there is for each n a real α_n such that

$\bar{\alpha}(n) = \bar{\alpha}_n(n)$ and $\bar{f}(n) = \langle\sigma_0(\alpha_n),\ldots,\sigma_{n-1}(\alpha_n)\rangle$.

But then $\langle\alpha_i \mid i<\omega\rangle$ converges to α and converges in norm to f. So $\alpha \in A$.

Notes:

1) If T is the tree associated with the scale $\vec{\sigma}$, then for $\alpha \in A$, $\vec{\sigma}(\alpha)$ is the leftmost branch through $T(\alpha)$.

2) Say that $\vec{\sigma}$ is a κ-scale if each of its norms has length less than or equal to κ. We've shown that A admits a κ-scale if and only if A is κ-Souslin. This is a little misleading because it blots out all considerations of definability.

3) We've made no use so far of semi-continuity. An essential application will eventually appear. (Notice that if we only wanted the closure property we could have chosen $\vec{\sigma}(\alpha)$ to be any branch through $T(\alpha)$.)

B. Digression. This section is not necessary to what follows. It discusses a few obvious ways of varying the definition of "scale".

Let $f \leq_p g$ mean that for all n, $f(n) \leq g(n)$. Say that $\vec{\sigma}$ is a pointwise scale if in the statement of the limit property we replace (2) by "$\vec{\sigma}(\alpha) \leq_p \vec{\lambda}$".

Proposition (i)

From any ξ-scale on A we can "easily" obtain a pointwise scale on A all of whose norms have length $\leq \xi^\omega$ (ordinal exponentiation).

Proof

Let $\langle\ ,\ \rangle$ be such that for any n, $\langle\ ,\ \rangle : {}^n\xi \longleftrightarrow \xi^n$ is an order isomorphism, where the set ${}^n\xi$ of sequences from ξ of length n is ordered by first point of difference (i.e., by $\leq_{lex}$). For any $f: \omega \to \xi$ let $f': \omega \to \xi^\omega$ be given by $f'(n) = \langle \bar{f}(n) \rangle$. Then $f \leq_{lex} g$ if and only if $f' \leq_p g'$. If $\vec{\sigma}$ is a scale on A then the sequence $\vec{\tau}$ defined by: $\tau_n(\alpha) = (\vec{\sigma}(\alpha))'(n)$ is a pointwise scale on A.

Proposition (ii)

The scale $\vec{\tau}$ obtained in the proof of proposition (i) is refined: For all m,n($n \leq m$ and $\sigma_m(\alpha) \leq \sigma_m(\beta) \Longrightarrow \sigma_n(\alpha) \leq \sigma_n(\beta)$).

Notes:

1) The word "easily" in proposition (i) means that $\vec{\tau}$ is a sufficiently mild perturbation of $\vec{\sigma}$ that if $\vec{\sigma}$ is a Γ scale (see section 3.C) then so, for any reasonable Γ, is $\vec{\tau}$.

2) The jump in ordinal from ξ to ξ^ω is significant: All $\underset{\sim}{\Sigma}^1_1$ sets have ω-scales, but only $\underset{\sim}{\Delta}^1_1$ sets have pointwise ω-scales.

3) Strengthening semicontinuity to continuity - i.e., $\vec{\sigma}(\alpha) = \vec{\lambda}$ - produces a properly stronger notion (called a strong scale).

With one more small perturbation we can obtain a very convenient sort of scale for which norm convergence implies topological convergence.

Definition 3.4

A scale $\vec{\nu}$ on A is nice iff

1) $\vec{\nu}$ is a refined, pointwise scale;

2) whenever a sequence from A converges in norm it converges topologically;

3) whenever two sequences from A converge in norm to the same sequence of ordinals they converge topologically to the same real.

Proposition (iii)

Any scale on A can be gently perturbed to a nice scale on A.

Proof

Use the previous trick, at the same time shuffling in α: Given $\vec{\sigma}$, let $\nu_n(\alpha) = \langle \alpha, \vec{\alpha}(\alpha) \rangle'(2n)\rangle$

C. Definable scales; the uniformization theorem.

Definition 3.5

$\vec{\sigma}$ is a Γ scale on A iff $\vec{\sigma}$ is a scale on A whose norms are uniformly Γ norms; i.e.

There exist subsets $\leq^{\Gamma}_{\smile}$ and $\leq^{\check{\Gamma}}$ of $\omega \times X \times X$ which are, respectively, elements of Γ and $\check{\Gamma}$, such that

For all $y \in A$, all $n \in \omega$, and all x,

$x \in A$ and $\sigma_n(x) \leq \sigma_n(y) \iff (n,x,y) \in \leq^{\Gamma}_{\smile}$

$\iff (n,x,y) \in \leq^{\check{\Gamma}}$

Definition 3.6

Scale $(\Gamma) =_{df}$ Every set in Γ has a Γ scale.

Fact (proof deferred): Scale (Π^1_1) and Scale (Σ^1_2).

Granted the Fact we'll employ Scale (Π^1_1) to prove the well known basis theorem:

Theorem 3.7

Every nonempty Π^1_1 subset of $\mathbb{R}$ contains a Π^1_1 singleton.

Proof

Let A be Π^1_1, $\vec{\sigma}$ a Π^1_1-scale on A, T its associated tree and α^* the canonically chosen element of p[T]. We'll show that $\{\alpha^*\}$ is a Π^1_1 set. Go back to the "internal" definition of α^*, marked (+) on page 155. We want to be able to rephrase that definition so as to refer (in a Π^1_1 way) only to reals, not to both reals and ordinals. The key is the semi-continuity property, which guarantees that, in the notation of (+), $(\xi_0, \xi_1, \ldots) = \vec{\sigma}(\alpha^*)$. (Proof: Note (1) to lemma 3.2 - which is a consequence of semi-continuity - points out that $\vec{\sigma}(\alpha)$ is the leftmost branch of $T(\alpha)$. But by construction $\vec{\xi}$ is the leftmost branch of $T(\alpha)$.) Let P(x) be the following predicate:

$x(0), x(1), \ldots$ are respectively $k_0, k_1, \ldots$
and $\sigma_0(x), \sigma_1(x), \ldots$ are $\xi_0, \xi_1, \ldots$

Then α^* is the unique solution to P. The check that "$x \in P$" is Π^1_1 is tedious, but routine. Here it is. (We're using "$\alpha \equiv_n \beta$" to abbreviate "$\sigma_n(\alpha) = \sigma_n(\beta)$", etc.)

Paraphrase of P(x)	Translation
$x \in A$ and	
$\forall\gamma \; \neg(\gamma <^{\Sigma}_{0} x)$	$\sigma_0(x) = \xi_0$
and	
$\forall\gamma \; (\gamma \equiv^{\Sigma}_{0} x \Longrightarrow x(0) \le \gamma(0))$	$x(0) = k_0$
and	
$\forall n \forall \gamma$ { $[\forall i \le n \; (\gamma \equiv^{\Sigma}_{i} x$ and $\gamma(i) = x(i))] \Longrightarrow [x \le^{\Pi}_{n+1} \gamma]$	$\sigma_{n+1}(x) = \xi_{n+1}$
and	
$\forall i \le n+1 (\gamma \equiv^{\Sigma}_{i} x$ and) and $\forall i \le n \; \gamma(i) = x(i)] \Longrightarrow x(n+1) \le \gamma(n+1)$ }	$x(n+1) = k_{n+1}$

Note: We've repeatedly used the fact that Π^1_1 is closed under $\forall^{\mathbb{R}}$.

Definition 3.8

Unif(Γ) $=_{df}$ For every $A \in \Gamma$, $A \subseteq \mathcal{X}\times\mathcal{Y}$, there exists $B \in \Gamma$, $B \subseteq \mathcal{X}\times\mathcal{Y}$ which is a selection function for A.

By relativizing the proof of theorem 3.6 we easily show that Unif(Π^1_1). Abstractly,

Theorem 3.9

Suppose that Γ is adequate and closed under $\forall^{\mathbb{R}}$. Then, Scale(Γ) $\Longrightarrow$ Unif(Γ).

Notice that 3.7 does not allow us to deduce Unif(Σ^1_2) from Scale (Σ^1_2). That is no problem since, trivially,

Theorem 3.10

Γ is adequate, Unif(Γ) $\Longrightarrow$ Unif($\exists^{\mathbb{R}}\Gamma$).

Corollary 3.11

Unif(Π^1_1) and Unif(Σ^1_2).

Note: It takes slightly more effort to demonstrate than the corresponding fact about pwo, but the scale property propogates from Γ to $\exists^{\mathbb{R}}\Gamma$ (when Γ is closed under $\forall^{\mathbb{R}}$). Hence, under V = L and PD the patterns for Scale (so, for Unif) are the same as those for PWO. These are not the only existence theorems. E.g., a theorem of Martin and Solovay says that if all the sharps exist, then every Π^1_2 set admits a Π^1_3 scale (of length u_ω = the ωth uniform indiscernible).

We'll conclude with a proof of Scale (Π^1_1).

Theorem 3.12 (Kondo-Novikoff-Addison-Moschovakis)

Scale (Π^1_1)

Proof

Let A be Π^1_1 and $f: A \to \mathbb{R}$ a recursive function such that $x \in A \iff f(x) \in$ WO. For any $n \in \omega$, put:

$$x \lesssim_n y \iff |f(x)| < |f(y)| \text{ or } (|f(x)| = |f(y)| \text{ and } |f(x){\restriction}n| \le |f(y){\restriction}n|)$$

where for any β, $\beta{\restriction}n = \{(i,j) \mid i \le_\beta j <_\beta n\}$.

(Equivalently, letting < , > again be the lex-order isomorphism $^2\omega \longleftrightarrow \omega^2$, $\sigma_n(x) = \langle |f(x)|, |f(x){\restriction}n| \rangle$.) It's left as an exercise to check that the σ_n are, uniformly, Π^1_1 norms. (Had we defined $\sigma_n(x)$ to be simply $|f(x){\restriction}n|$ it wouldn't follow that each σ_n is a Π^1_1 norm.)

Checking the limit property: Suppose that $\langle x_i \mid i<\omega \rangle$ is a sequence from A converging to x and converging in norm to $\vec{\lambda}$. It must be the case that as $i \to \infty$, the $|f(x_i)|$ become eventually constant with value, say, η. So, using the alternative definition of the σ_n's, there is for each n a unique ordinal λ'_n with $\lambda_n = \langle \eta, \lambda'_n \rangle$.

Closure property: We'll show $f(x) \in$ WO. It will suffice to show that $m <_{f(x)} n \implies \lambda'_m < \lambda'_n$. So suppose $m <_{f(x)} n$. Since the $f(x_i)$ converge to $f(x)$ we can choose an i large enough to insure that $m <_{f(x_i)} n$, $|f(x_i){\restriction}m| = \lambda'_m$ and $|f(x_i){\restriction}n| = \lambda'_n$. But to say that $m <_{f(x_i)} n$ is to say precisely that $|f(x_i){\restriction}m| < |f(x)_i{\restriction}n|$.

Semi-continuity: More or less by definition, (a) $|f(x)| = \sup\{|f(x){\restriction}n| + 1 : n \in \omega\}$. We also have (b) $\eta \ge \sup\{\lambda'_{n+1} \mid n \in \omega\}$;

for, if i is large enough, $|f(x_i)| = \eta$ and $|f(x_i)\restriction n| = \lambda'_n$. Since we've just shown that the map $n \to \lambda'_n$ is order preserving on $\leq_{f(x)}$, (c) $|f(x)\restriction n| \leq \lambda'_n$ for each n. Putting (a) - (c) together we obtain: $f(x) \leq \eta$. Using this and, once again, fact (b) we have $\sigma_n(x) = <|f(x)|,|f(x)\restriction n|> \leq <\eta,\lambda'_n> = \lambda_n$ for each n. So $\vec{\sigma}$ is in fact a pointwise scale on A.

D. The reader who would like to see trees, norms, and scales applied in a variety of ways will find

Kechris, A.S. "The theory of countable analytical sets" TAMS 202, (1975) 259-297

Kechris, A.S. "Measure and category in effective descriptive set theory", Annals Math. Log. 5, (1973) 337-384

interesting and rich (and will find further references as well).

On the regularity of ultrafilters

Karel Prikry*

It was shown in [6] that in Gödel's constructible universe L, every uniform ultrafilter over ω_1 is regular. This involved a new combinatorial principle stronger that Kurepa's hypothesis. Chang [2] and Jensen [4] have generalized this principle to higher cardinals and extended the regularity result of [6] to all ω_n, $(n \in \omega)$. Benda and Ketonen [1] have generalized and simplified the result of [6] by making use of a weaker combinatorial principle. In the present note we show how to generalize the Chang-Jensen result along similar lines. Kunen asked in a discussion whether this can be done.

We start by formulating the transversal hypothesis in a customary form:

$TH(\lambda,\nu)$: There is a family $F \subseteq {}^{\lambda}\nu$ such that $|F| = \lambda^+$ and for all $f,g \in F$, if $f \neq g$, then $f(\gamma) \neq g(\gamma)$ for all sufficiently large $\gamma \in \lambda$.

The following consistency results are well known.

<u>Theorem 1</u>, (Solovay, see [3]). In L, $TH(\kappa^+,\kappa)$ holds for all infinite cardinals κ.

In the opposite direction we only give a special case of Silver's Theorem [7].

<u>Theorem 2</u>. If ZFC + there is a Ramsey cardinal is consistent, then ZFC + $\neg TH(\omega_1,\omega)$ is consistent.

* The author received support from the NSF Grant MCS 74-06705.

For regular λ it follows trivially from $TH(\lambda,\nu)$ that $\lambda \leq \nu^+$. An obvious attempt at generalizing the Benda-Ketonen approach to, say ω_2, leads to considering statements like $TH(\omega_2,\omega)$. But there does not exist a family of even ω_1 almost disjoint functions from ω_2 into ω, let alone ω_3 such functions.

Definition 1. A uniform ultrafilter U over λ is (λ,ν)-regular if there are sets $a_\alpha \in U$, $(\alpha\in\lambda)$, such that for all $I \in [\lambda]^\nu$, $\cap\{a_\alpha : \alpha\in I\} = 0$. U is regular if U is (λ,ω)-regular.

Benda and Ketonen showed

Theorem 4, [1]. If $TH(\kappa^+,\kappa)$ holds, then every uniform ultrafilter over κ^+ is (κ^+,κ)-regular.

As a special case, if $TH(\omega_1,\omega)$ holds, then every uniform ultrafilter over ω_1 is regular, which is an optimal result. By Theorem 1, this holds in L as well.

We shall now introduce a modified form of the transversal hypothesis:

$MTH(\kappa,\nu)$: There is a family F such that each $f \in F$ is a function $f : [\kappa]^{\leq\nu} \to \nu$, $|F| = \kappa^+$, and for every $f \in F$, and every $G \subseteq F$, if $|G| < \kappa$ and $f \notin G$, then there is $\alpha \in \kappa$ such that $f(x) \neq g(x)$ for all $g \in G$ and all $x \in [\kappa]^{\leq\nu}$ satisfying $\alpha \in x$.

Remark. Note that if we require only $|F| = \kappa$, rather than κ^+, in the formulation of MTH, then we can prove the resulting statement for all infinite κ and ν. If we contrast this situation with the preceding remark concerning $TH(\omega_2,\omega)$, it becomes clear that formulating MTH is a step in the right direction. We proceed to prove the claim in this remark. This is non-trivial only when $\nu < \kappa$. For each $x \in [\kappa]^{\leq\nu}$ let g_x be an injection of x into $\nu - \{0\}$. Now for each $\alpha \in \kappa$, $\alpha \neq 0$, define

$$f_\alpha(x) = \begin{cases} g_x(\alpha) & \text{if } \alpha \in x \\ 0 & \text{if } \alpha \notin x \end{cases}$$

Let $\alpha,\beta \in \kappa$, $\alpha \neq \beta$, $\alpha,\beta \neq 0$. Then if $\alpha \in x$, $f_\alpha(x) \neq f_\beta(x)$.

Our main result is

Theorem 5. If $MTH(\kappa,\nu)$ holds, then every uniform (κ,ν^+)-regular ultrafilter over κ is (κ,ν)-regular.

We shall first of all derive some corollaries, in particular the Chang-Jensen result. Kurepa's hypothesis, $KH(\kappa,\nu)$ is the statement:

There is a family $S \subseteq [\kappa]^\kappa$ such that $|S| = \kappa^+$ and for all $x \in [\kappa]^\nu$, $|S \restriction x| \leq \nu$, where $S \restriction x = \{s \cap x : s \in S\}$.

Let $KH'(\kappa,\nu)$ be $KH(\kappa,\nu)$ + "S is κ-almost disjoint".

A slight modification of Jensen's proof of $KH(\kappa,\nu)$ gives $KH'(\kappa,\nu)$. More precisely,

Theorem 6. $KH'(\kappa,\nu)$ holds in L whenever κ is regular, not ineffable, and $\nu < \kappa$.

Proposition 1:

For regular κ, $KH'(\kappa,\nu)$ implies $MTH(\kappa,\nu)$.

Proof: Let $S \subseteq [\kappa]^\kappa$ be as in $KH'(\kappa,\nu)$. Now define for each $s \in S$, $f_s(x) = s \cap x$, $(x \in [\kappa]^{\leq\nu})$. Since there are only ν sets of the form $x \cap s$ where s ranges over S, we can regard f_s as a function into ν. We set $F = \{f_s : s \in S\}$. Let $G \subseteq F$, $|G| < \kappa$, $f_{s_0} \in F-G$. Let $R = \{r \in S : f_r \in G\}$. Let $\alpha \in s_0 - R$. Then f_{s_0} disagrees with all $g \in G$ on those x which contain α.

The Chang-Jensen result now follows from Theorems 5, 6 and Proposition 1:

Theorem 7. The following holds in L: If κ is regular, not ineffable and $\nu < \kappa$, then every uniform (κ,ν^+)-regular ultrafilter over κ is (κ,ν)-regular.

Corollary. In L, every uniform ultrafilter over ω_n is regular, $(n \in \omega)$.

Proof of Theorem 5:

Let $a_\alpha \in U$, $(\alpha \in \kappa)$, $\bigcap\{a_\alpha : \alpha \in I\} = 0$ if $|I| = \nu^+$. As usual, we find $\tilde{U} \leq_{RK} U$ where $\tilde{U}$ is over $[\kappa]^{\leq\nu}$. Namely, define $h : \kappa \to [\kappa]^{\leq\nu}$ by $h(\rho) = \{\alpha \in \kappa : \rho \in a_\alpha\}$. Hence $|h(\rho)| \leq \nu$. We set $\tilde{U} = h^*(U)$. Then for each $\alpha \in \kappa$, $r_\alpha = \{x \in [\kappa]^{\leq\nu} : \alpha \in x\} \in \tilde{U}$. It now suffices to show that $\tilde{U}$ is (κ,ν)-regular.

Let F be as in MTH(κ,ν). If f, g $\in$ F and f $\neq$ g, then f,g differ on some r_α, and thus mod $\tilde{U}$.

Let $\prec$ be the ultraproduct ordering on F. Since $|F| = \kappa^+$, there is some $\bar{f} \in F$ which has κ predecessors in $\prec$. Let them be f_ξ, $(\xi\in\kappa)$, f_ξ distinct.

We can now define the (κ,ν)-regularising family for $\tilde{U}$. Suppose that $b_\xi \in \tilde{U}$, $(\xi < \eta)$, are defined, where $\eta < \kappa$. Consider the family $\{f_\xi : \xi < \eta\} \subseteq F$, and f_η - not in the family. By MTH(κ,ν) we can fix $\alpha_\eta \in \kappa$ such that $f_\eta(x) \neq f_\xi(x)$ for all $x \in r_{\alpha_\eta}$, all $\xi < \eta$. Set $b_\eta = r_{\alpha_\eta} \cap \{x\in[\kappa]^{\leq\nu} : f_\eta(x) < \bar{f}(x)\}$.

We claim that b_ξ, $(\xi<\kappa)$, is (κ,ν)-regularising for $\tilde{U}$. Suppose not and let $I \subseteq \kappa$ be such that $|I| = \nu$, and $\cap\{b_\xi : \xi\in I\} \neq 0$. Pick x_0 in this intersection. It then follows that $f_\xi(x_0)$, $(\xi\in I)$, are all distinct. Indeed, if $\xi < \eta$, $\xi, \eta \in I$, then $x_0 \in r_{\alpha_\eta}$, and $f_\eta(x_0) \neq f_\xi(x_0)$ follows from the choice of α_η. On the other hand, $f_\xi(x_0) < \bar{f}(x_0) < \nu$ for all $\xi \in I$, and $|I| = \nu$. This contradition completes the proof of Theorem 5.

University of Minnesota
Minneapolis, Minnesota.

REFERENCES

1. M. Benda and J. Ketonen, On regularity of ultrafilters, Israel Journ. of Math. 17 (1974), 231-240.

2. C.C. Chang, Extensions of combinatorial principles of Kurepa and Prikry, Theory of sets and topology, ed. G. Asser et. al., Berlin 1972.

3. Keith J. Devlin, Aspects of constructibility, Springer-Verlag, New York 1973.

4. R. Jensen, Some combinatorial principles in L and V, an unpublished manuscript.

5. Richard Laver, A set in L containing regularizing families for ultrafilters. Mathematika 24 (1977), 50-51.

6. K. Prikry, On a problem of Gilman and Keisler, Annals Math. Logic 2 (1970), 179-187.

7. Jack Silver, The independence of Kurepa's conjecture and two cardinal conjectures in model theory, Axiomatic Set Theory, Proc. Symp. Pure Math., Vol.XIII, Part I, ed. D. Scott, Amer. Math. Soc., Providence 1971, p. 383-390.

MORASSES IN COMBINATORIAL SET THEORY

Akihiro Kanamori

Baruch College
City University of New York
New York, NY 10010

Jensen invented the morass in order to establish strong model-theoretic transfer principles in the constructible universe. Morasses are structures of considerable complexity, a culminating edifice in Jensen's remarkable program of formulating useful combinatorial principles which obtain in the constructible universe, and which moreover can be appended to any model of set theory by straightforward forcing. Gödel's Axiom of Constructibility V = L is surely the ultimate combinatorial principle in ZFC, and the morass codifies a substantial portion of the structure of L. As set theorists looked beyond the well-known $\Diamond$ and $\Box$ for applicable combinatorial principles, it was natural to consider extractions from the full structure of a morass.

This paper is an expository survey that schematizes the higher combinatorial principles derivable from morasses which have emerged in set theoretical praxis. It is notable that most of these principles were formulated by combinatorial set theorists to isolate salient features of particular constructions, and shown by them to be consistent first by forcing. Then, the specialists in L established how they hold there, in ad hoc fashion using the full structure of the morass. The sections of this paper deal successively with Prikry's Principle, Silver's Principle, Burgess' Principle, and finally limit cardinal versions. This sequence reflects the historical development of ideas, the progression toward further complexity, and coincidentally the author's series of papers [Ka2][Ka3] [Ka4]. The cumulative layers of sophistication provide an illuminating approach to the full morass structure, whilst at the same time providing a hierarchy of principles which, seen in this scheme, will hopefully find wider application in the future. The emphasis will be on shorter, illustrative proofs for the casual but interested reader, with adequate references for the more persistent researcher.

The set theoretical notation is standard, and here is a short litany: The first Greek letters

$\alpha, \beta, \gamma, \ldots$ denote ordinals, whereas the middle Greek letters $\kappa, \lambda, \mu, \ldots$ are reserved for infinite cardinals. If x is a set, $|x|$ denotes its cardinality, $[x]^{\kappa}$ denotes the collection of subsets of x of cardinality κ, and if f is a function, $f''x = \{f(y) \mid y \in x\}$. Finally, ${}^{y}x$ denotes the set of functions from y into x.

§1. PRIKRY'S PRINCIPLE

In the first three sections, κ will always denote a successor cardinal, with κ^- its predecessor. The general situation will be considered in the last section. Prikry's Principle is the following proposition:

(P_κ) There is a collection $\{f_\alpha | \alpha < \kappa^+\} \subseteq {}^\kappa\kappa$ so that whenever $s \in [\kappa^+]^{\kappa^-}$ and $\phi \in {}^s\kappa$, we have $|\{\xi < \kappa | \forall\alpha\in s(f_\alpha(\xi) \neq \phi(\alpha))\}| < \kappa$.

This first approach to the system of approximations which comprises a morass says roughly that there are κ^+ functions: $\kappa \to \kappa$ such that: if guesses are made at possible values for any κ^- many of them, then for sufficiently large $\xi < \kappa$, at least one guess is rendered correct at ξ. Although it is not made explicit, notice that we can assume that the f_α's are pairwise distinct and have range $= \kappa$, since easy applications of P_κ show that only $< \kappa^-$ of these functions do not have these properties. Historically, P_κ was the first of the higher combinatorial principles to be formulated In a ground-breaking paper, Prikry [P] devised his principle and established its consistency with the GCH by a method of forcing with side conditions. Assuming $2^{\kappa^-} = \kappa$, a simple diagonal argument provides κ functions satisfying the conclusion of P_κ; Prikry's argument yields κ^+ many, and in fact can provide arbitrarily many in a cardinal-preserving forcing extension. There was no particular emphasis laid on possibilities in L, but at any rate, the first result about L in this whole context was due to Jensen, who showed that if V = L, then P_κ holds for every successor cardinal κ, using the morass structure that he invented.

Prikry was answering a question of Erdős, Hajnal and Rado [EHR] in the partition calculus, and this was the first example of the phenomenon of relative consistency results, rather than outright demonstrations, in this area of set-theoretical research. To recall the relevant case of the polarized partition symbol of [EHR],

$$\begin{pmatrix}\lambda\\ \kappa\end{pmatrix} \longrightarrow \begin{bmatrix}\mu\\ \nu\end{bmatrix}_\gamma$$

means that whenever $F\colon \lambda\times\kappa \to \gamma$, there are $X \in [\lambda]^\mu$ and $Y \in [\kappa]^\nu$ such that $F''(X\times Y) \neq \gamma$. To denote the negation of this proposition, $\to$ is replaced by $\not\to$. Besides [EHR], see the secondary source Williams [Wi] Chapter 4 for background. Prikry educed P_κ from the negative partition relation that he wanted to show consistent, which is equivalent to the ostensibly weaker version where the $\phi \in {}^S\kappa$ just range over the constant functions:

$$P_\kappa \to \binom{\kappa^+}{\kappa} \not\to \begin{bmatrix}\kappa^- \\ \kappa\end{bmatrix}_\kappa$$

(If $\{f_\alpha \mid \alpha < \kappa^+\}$ is as provided by P_κ, set $F(\alpha,\beta) = f_\alpha(\beta)$ to get a counter-example.)

P_κ has several other applications. Prikry himself in [P] provided a consequence about an old problem of Ulam's on measurability with respect to a sequence of measures. In a related development, Szymański [Sz] formulated the following concept in order to establish some Baire Category-type theorems for $U(\omega_1)$, the space of uniform ultrafilters over ω_1. For any infinite cardinal λ, a matrix $\{A^n_\alpha \mid n < \omega,\ \alpha < \lambda\}$ is a <u>λ-matrix</u> <u>iff</u>

(a) if $m < n$ and $\alpha < \lambda$, then $A^m_\alpha \subseteq A^n_\alpha$,

(b) $\bigcup\{A^n_\alpha \mid n < \omega\} = \omega_1$ for each $\alpha < \lambda$, and

(c) for every infinite $s \subseteq \lambda$ and $\phi \in {}^s\omega$,

$$|\bigcap\{A^{\phi(\alpha)}_\alpha \mid \alpha \in s\}| < \omega_1.$$

A basic clopen set for $U(\omega_1)$ is a set of form $\{u \in U(\omega_1) \mid A \in u\}$ for some $A \subseteq \omega_1$; and a G_δ closed set is a countable intersection of basic clopen sets. Szymański established the following equivalence: A λ-matrix exists <u>iff</u> there is a family of λ G_δ closed and nowhere dense subsets of $U(\omega_1)$ such that the union of any infinite subfamily is dense in $U(\omega_1)$. The connection with P_{ω_1} is clear:

If P_{ω_1}, then there is an ω_2-matrix.

(Let $g: \omega_1 \to \omega$ be any surjection, and given $\{f_\alpha \mid \alpha < \omega_2\}$ as provided by P_{ω_1} set $A^n_\alpha = \{\xi \mid g \cdot f_\alpha(\xi) \leq n\}$.) Incidentally, Baumgartner has established that the following are equivalent (see [Ka2] for a proof): (a) an ω_1-matrix exists; (b) an ω-matrix exists; and (c) there is a subset of ${}^\omega\omega$ of cardinality ω_1 without an upper bound in ${}^\omega\omega$ under the ordering of eventual dominance.

Delimiting the ranges of the functions in P_κ even further by composing with a fixed surjection $g: \kappa \to 2$ yields the following:

(P^-_κ) There is a collection $\{f_\alpha \mid \alpha < \kappa^+\} \subseteq {}^\kappa 2$ so that whenever $s \in [\kappa^+]^{\kappa^-}$ and $\phi \in {}^s 2$, we have $|\{\xi < \kappa \mid \forall \alpha \in s (f_\alpha(\xi) \neq \phi(\alpha))\}| < \kappa$.

Even this weakened principle has its uses: Balcar, Simon and Vojtáš asked([BSV] Problem 20b) whether the following is consistent: whenever λ is regular and uncountable and U is a uniform ultrafilter over λ, then there are λ^+ sets in U such that the intersection of any infinitely many of them has cardinality $< \lambda$. Probably, this is true in L, and the proof will depend heavily on the structure of ultrafilters. But at least, one can affirm the case $\lambda = \omega_1$: If P^-_κ, then for any uniform ultrafilter over κ there are κ^+ sets in U such that any κ^- of them has intersection of cardinality $< \kappa$. (The proof is immediate.)

Perhaps a more substantial application of P_κ is to the <u>Hajnal-Máté Principle</u>, formulated in the study of set mappings in combinatorial set theory, in Hajnal-Máté [HM]:

(HM_κ) There is a collection $\{h_\xi \mid \xi < \kappa\} \subseteq {}^{\kappa^+}\kappa^+$ with $h_\xi(\alpha) \neq \alpha$ for every $\xi < \kappa$ and $\alpha < \kappa^+$ such that: whenever $s \in [\kappa^+]^{\kappa^-}$, we have $|\{\xi < \kappa \mid \forall \alpha \in s(h_\xi(\alpha) \notin s)\}| < \kappa$.

In the most general setting, a <u>set mapping</u> on a set X is a function f from a subset of P(X) into P(X) such that $x \cap f(x) = \emptyset$ for every x in the domain. A subset $H \subseteq X$ is <u>free with respect to f</u> <u>iff</u> $H \cap f(x) = \emptyset$ for every $x \subseteq H$ in the domain. The general problem of when large free subsets for set mappings exist was extensively investigated through the fifties by classical combinatorial means in Eastern Europe. (See the secondary source Williams [Wi] Chapter 3 for background; a timely application of this theory is found in Galvin-Hajnal [GH].) Forcing and in particular Prikry's method of forcing with side conditions extended the realms of possibility, and Hajnal and Máté distilled their principle with the following implication in mind: If HM_κ and there is a κ-Kurepa tree (see the next section), then there is an $f: [\kappa^+]^3 \to \kappa^+$ such that no set of cardinality κ^- is free with respect to f. Note that HM_κ itself is a proposition about set mappings and free sets. Fitting into the pattern, Hajnal and Máté established the consistency of HM_κ by forcing, and then it was shown later to be true in L, this time by Burgess [Bu2], who first established a stronger principle in L to be discussed in §3.

<u>Theorem 1 (Kanamori)</u>: $P_\kappa \to HM_\kappa$.

$\vdash$ Suppose that $\{f_\alpha | \alpha < \kappa^+\}$ is as provided by P_κ. For each $\alpha < \kappa^+$, let $\psi_\alpha: \alpha \to \kappa$ be injective. Finally, define $h_\xi: \kappa^+ \to \kappa^+$ for $\xi < \kappa$ by:

$$h_\xi(\alpha) = \begin{cases} 1 & \text{if } \alpha = 0, \text{ else} \\ \psi_\alpha^{-1}(f_\alpha(\xi)) & \text{if this is defined, and} \\ 0 & \text{otherwise.} \end{cases}$$

Thus, we took care that $h_\xi(\alpha) \neq \alpha$ for every $\alpha < \kappa^+$.

To verify HM_κ, suppose that $s \in [\kappa^+]^{\kappa^-}$. Let β be the least element in $s - \{0,1\}$, and set $t = s - (\beta+1)$. Define $\phi: t \to \kappa$ by $\phi(\alpha) = \psi_\alpha(\beta)$. Then

$\{\xi < \kappa | \forall\alpha\in t(f_\alpha(\xi) \neq \phi(\alpha))\} \supseteq \{\xi < \kappa | \forall\alpha\in s(h_\xi(\alpha) \notin s)\}$

and hence by P_κ, this last set has cardinality $< \kappa$. $\dashv$

§2. SILVER'S PRINCIPLE

In order to formulate Silver's Principle (and also Burgess' Principle in the next section), it will be helpful to establish once and for all some conventions regarding trees. If T is a tree, T_ξ will denote the members of T at its ξth level; if $x \in T_\zeta$ and $\xi \leq \zeta$, then $\pi_\xi(x)$ is the tree predecessor of x at level ξ. Let us assume that trees are normalized at limits, i.e. if δ is a limit ordinal and $x \neq y$ are both in the δth level, then there is a $\xi < \delta$ such that $\pi_\xi(x) \neq \pi_\xi(y)$. A <u>$\kappa$-Kurepa tree</u> is a tree with height $\kappa+1$ such that $|T_\kappa| > \kappa$, yet $|T_\xi| \leq \xi$ for every $\xi < \kappa$. (This is congruous with the usual definition; it will be convenient to identify cofinal branches with a top level.) This settled, here is Silver's Principle:

(W_κ) There is a κ-Kurepa tree T and a function W with domain κ such that:

(a) for each $\xi < \kappa$, we have $W(\xi) \subseteq [T_\xi]^{\kappa^-}$ with $|W(\xi)| \leq \kappa^-$.

(b) for any $s \in [T_\kappa]^{\kappa^-}$, there is a $\gamma < \kappa$ such that whenever $\gamma \leq \xi < \kappa$, we have π_ξ"$s \in W(\xi)$.

Like P_κ, W_κ meets requirements mandated by κ^- size subsets of κ^+ (in essence, as $|T_\kappa| \geq \kappa^+$) at all sufficiently large stages. The new feature in the ascent towards the morass is the κ-Kurepa tree structure: the system of approximations has small initial stages. This combines with the potency provided by the function W, which plays a role somewhat akin to the sequence of distinguished subsets in $\diamondsuit^+$. Notice that if T is any tree of height $\kappa+1$, even if T were not necessarily a κ-Kurepa tree, as long as $|T_\kappa| \geq \kappa^-$ and there is a function W satisfying (a) and (b) above for this T, it is not difficult to see that $2^{\kappa^-} = \kappa$ must be satisfied.

The formulation of W_κ evinced an evolution from P_κ in the focus of attention as well. Upon seeing some consistency results constructed by Hajnal and Juhász in set-theoretic topology, Silver extracted W_κ from the morass in order to effect these constructions in L. So, unlike P_κ, W_κ was formulated with ramifications in L in mind. That many, long-winded combinatorial emanations from a morass actually follow from W_κ is a testimonial to Silver's insight. Incidentally, the consistency argument for W_κ through forcing is not difficult. To the usual notion of forcing for adjoining a κ-Kurepa tree one appends further clauses reminiscent of Prikry's side conditions (see Burgess [Bu1] for an exposition). The following proof illustrates the constructions possible with W_κ:

Theorem 2 (Silver): $W_\kappa \to P_\kappa$.

$\vdash$ Let T and W be as provided by W_κ; we might as well assume that $T_\kappa = \kappa^+$ by renaming. Since W_κ implies $2^{\kappa^-} = \kappa$, for each $\delta < \kappa$ and $s \in W(\delta)$, we can enumerate ${}^s\kappa$ as $\{h^s_\xi \mid \delta \leq \xi < \kappa\}$. To take care of more and more of the h^s_ξ's, for each $\xi < \kappa$ we shall define functions $g_\xi\colon T_\xi \to \kappa$ such that:

(†) Whenever $\delta \leq \eta \leq \xi$, $s \in W(\xi)$, and π_δ"$s \in W(\delta)$, there is an $x \in s$ such that $g_\xi(x) = h^{\pi_\delta\text{"}s}_\eta(\pi_\delta(x))$.

Once this is done, the proof can be completed by defining $f_\alpha\colon \kappa \to \kappa$ for $\alpha < \kappa^+$ by:

$$f_\alpha(\xi) = g_\xi(\pi_\xi(\alpha)).$$

To verify P_κ, let $s \in [\kappa^+]^{\kappa^-}$ and $\phi \in {}^s\kappa$. By hypothesis, there is a $\delta < \kappa$ such that $\delta \leq \xi < \kappa$ implies π_ξ"$s \in W(\xi)$, where we can take δ sufficiently large so that π_δ is injective on s. For some $\eta \geq \delta$, we have $h^{\pi_\delta\text{"}s}_\eta = \phi \cdot \pi_\delta^{-1}$. (Here, π_δ^{-1} is clear from the context as that inverse of π_δ whose range is the top level.) Then for any ξ such that $\eta \leq \xi < \kappa$, there is an $x \in \pi_\xi$"s such that $g_\xi(x) = h^{\pi_\delta\text{"}s}_\eta(\pi_\delta(x))$. But if $\alpha = \pi_\xi^{-1}(x)$, then $f_\alpha(\xi) = g_\xi(x) = \phi(\alpha)$.

All that remains is to define the functions g_ξ so as to satisfy (†). But doing this is easy: Fix $\xi < \kappa$, and let $\{\langle s_\zeta, \delta_\zeta, \eta_\zeta\rangle \mid \zeta < \kappa^-\}$ enumerate all triples $\langle s, \delta, \eta\rangle$ where $\delta \le \eta \le \xi$, $s \in W(\xi)$, and π_δ"$s \in W(\delta)$. Now define exactly one value for g_ξ in each of κ^- stages inductively: If $\zeta < \kappa^-$, since only ζ values have been determined before the ζth stage and s_ζ has cardinality κ^-, there is an $x \in s_\zeta$ such that $g_\xi(x)$ has not yet been defined. Set $g_\xi(x) = h_\eta^{\pi_\delta\text{"}s}(\pi_\delta(x))$, where $\delta = \delta_\zeta$ and $\eta = \eta_\zeta$. Finally, after κ^- stages extend g_ξ arbitrarily to all of T_ξ. This completes the construction of g_ξ, and the proof is thus complete. $\dashv$

Let us now turn to the work of Hajnal and Juhász alluded to earlier. Pondering the existence of special topological spaces of large cardinality, Hajnal and Juhász realized in the early 1970's that concrete constructions readily follow from certain existential principles concerning matrices of sets. The following proposition is the strongest form of these principles, and can be appropriately dubbed the <u>Hajnal-Juhász Principle</u>:

(HJ_κ) There is a collection $\{f_\alpha \mid \alpha < \kappa^+\} \subseteq {}^\kappa 2$ so that whenever $\rho < \kappa^-$ and $s\colon \kappa^- \times \rho \to \kappa^+$ is injective, there is a $\gamma < \kappa$ such that: if $x \in [\kappa-\gamma]^{<\omega}$ and $\{\varepsilon_\tau \mid \tau < \rho\} \subseteq {}^x 2$, there is a $\sigma < \kappa^-$ with $\varepsilon_\tau \subseteq f_{s(\sigma,\tau)}$ for every $\tau < \rho$.

In set-theoretic topology, HL and HS are acronymic for hereditarily Lindelöf and hereditarily separable, respectively, and an L space is an HL space which is not HS, whilst an S space is an HS space which is not HL. There is quite a literature on the study of these spaces nowadays, particularly in connection with Martin's Axiom, and a good but older reference is M.E. Rudin [Ru] Chapter 5. An initial version of HJ_κ was considered by Hajnal and Juhász with the restriction to just $\rho = 1$. Taking the concrete case $\kappa = \omega_1$ (otherwise, we would have to frame the discussion in general terms around κ^--Lindelöf and κ^--separable), they show [HJ1] that this restricted principle implies the existence of normal S spaces of large cardinality, the so-called HFD spaces, and establish its consistency by forcing. Then Devlin [D] established this restricted principle in L, directly using morasses. As Hajnal and Juhász later realized, the full principle HJ_{ω_1} implies the existence of normal, strong S spaces of large cardinality. (A strong S space is a space X such that X^n is an S space for every $n \in \omega$.) Kunen [Ku2] has shown that under MA + $\neg$CH, there are no strong S spaces.

Concerning L spaces, Hajnal and Juhász early on [HJ2] formulated the following principle to construct

L spaces of large cardinality:

(HJ_κ^-) There is a collection $\{f_\alpha \mid \alpha < \kappa^+\} \subseteq {}^\kappa 2$ so that whenever $\rho < \kappa^-$ and $s\colon \kappa^- \times \rho \to \kappa^+$ is injective and $\phi\colon \kappa^- \times \rho \to 2$, we have

$$|\{\xi < \kappa \mid \forall\sigma<\kappa^- \exists\tau<\rho (f_{s(\sigma,\tau)}(\xi) \neq \phi(\sigma,\tau))\}| < \kappa.$$

(Actually, they had a further condition on $\{f_\alpha \mid \alpha < \kappa^+\}$ to insure good separation properties for the space constructed, but this is the crux of the matter.) HJ_κ^- immediately implies P_κ^-; that HJ_κ^- follows from HJ_κ is not unexpected, and details are provided in [Ka3]. The proof of the following theorem is also given in full in [Ka3]; it can be culled [HJ1][HJ2] and especially [Ju] where a simpler conclusion is derived.

Theorem 3 (Silver): $2^{<\kappa^-} = \kappa^-$ and $W_\kappa \to HJ_\kappa$.

§3. BURGESS' PRINCIPLE

Although we saw in §1 that the Hajnal-Máté Principle follows from Prikry's Principle, Burgess [Bu2] originally established the Hajnal-Máté Principle in L from a more complicated principle. The following, asserting the existence of what he calls quagmires, can be dubbed Burgess' Principle. Here, the notation for κ-Kurepa trees developed in §2 is still in effect.

(B_κ) There is a <u>quagmire</u>, i.e. a κ-Kurepa tree T with tree ordering $<$, equipped with a binary relation $\triangleleft$ and a ternary function Q such that:

(1) $y \triangleleft x$ implies that x and y are distinct elements on the same level, and $\triangleleft$ linearly orders every level.

(2) Q is defined on triples $\langle\bar{y},\bar{x},x\rangle$ just in case $\bar{y} \triangleleft \bar{x} < x$, and for any such, $y < Q(\bar{y},\bar{x},x) \triangleleft x$.

(3) (Commutativity) If $\bar{y} \triangleleft \bar{x} < x' < x$, then $Q(Q(\bar{y},\bar{x},x'),x',x) = Q(\bar{y},\bar{x},x)$.

(4) (Coherence) If $\bar{z} \triangleleft \bar{y} \triangleleft \bar{x} < x$, then $Q(\bar{z},\bar{y},Q(\bar{y},\bar{x},x)) = Q(\bar{z},\bar{x},x)$.

(5) (Completeness) If $y \triangleleft x \in T_\kappa$, then for some $\xi < \kappa$, $\pi_\xi(y) \triangleleft \pi_\xi(x)$ and $Q(\pi_\xi(y),\pi_\xi(x),x) = y$.

The reader familiar with morasses will already see a growing resemblance, and as with morasses, he or she is advised to draw pictures to get the picture. B_κ may seem a bit ad hoc, but it is really the next natural rung in the evolutionary ladder toward a morass. Whereas W_κ merely hypothesized κ^+ cofinal branches and a system of approximations by κ^- size subsets, B_κ endows a linear order on these branches which is moreover reflected in the $\triangleleft$ orderings through the previous levels. Thus, B_κ incorporates an important feature of morasses; the main ingredient which must still be added to get the full structure of a morass is the limit continuity across levels. Burgess [Bu2] established the following result:

<u>Theorem 4 (Burgess)</u>: If $2^{\kappa^-} = \kappa$ and B_κ, then W_κ.

$\vdash$ The idea here is first to enumerate the powerset $P(T_\xi)$ as $\{X_{\xi\rho} \mid \rho < \kappa\}$ for each $\xi < \kappa$, using $2^{\kappa^-} = \kappa$. For $x \in T_\gamma$ and $\xi, \rho < \gamma$ let $S(\xi,\rho,x) = \{Q(\bar{y},\pi_\xi(x),x) \mid \bar{y} \lhd \pi_\xi(x) \text{ and } \bar{y} \in X_{\xi\rho}\}$. Finally, for $\gamma < \kappa$ set $W(\gamma) = \{S(\xi,\rho,x) \mid \xi,\rho < \gamma \text{ and } x \in T_\gamma\}$. Then this function W works, capturing more and more of the images of the κ^- size subsets of the top level as we move to the right along $\lhd$ and upwards along $<$:

The Completeness condition (5) implies that any $x \in T_\kappa$ has at most κ $\lhd$-predecessors. Hence, given any $s \in [T_\kappa]^{\kappa^-}$, there is an $x \in T_\kappa$ such that $y \lhd x$ for every $y \in s$. Again by Completeness, for each $y \in s$ there is a $\xi(y) < \kappa$ such that $Q(\pi_{\xi(y)}(y),\pi_{\xi(y)}(x),x) = y$. Set $\delta = \sup\{\xi(y) \mid y \in s\}$. Then an easy application of Commutativity shows that for any ξ with $\delta \leq \xi < \kappa$ and $y \in s$, we have $\pi_\xi(y) \lhd \pi_\xi(x)$ and $Q(\pi_\xi(y),\pi_\xi(x),x) = y$. Finally, let ρ be such that $X_{\delta\rho} = \pi_\delta{}''s$. Then for any $\xi \geq \max(\delta,\rho)$, the above arguments confirm that $\pi_\xi{}''s = S(\delta,\rho,\pi_\xi(x)) \in W(\xi)$, completing the proof. $\dashv$

Just as Silver's Principle establishes Prikry's Principle, Burgess' Principle establishes an extended

Prikry's Principle, first formulated by Rebholz [Re] soon after morasses first saw the light of day. Rebholz' Principle is the following:

(R_κ) There is a collection $\{f_\alpha | \alpha < \kappa^+\}$ of functions with $f_\alpha: \alpha \to \alpha$, so that whenever $s \in [\kappa^+]^{\kappa^-}$ and ϕ is a regressive function with domain = s (i.e. $\phi(\alpha) < \alpha$ for $\alpha \in s$), then $|\{\xi < \bigcap s | \forall\alpha\in s(f_\alpha(\xi) \neq \phi(\alpha))\}| < \kappa$.

Using morasses (and $\Diamond_\kappa$, although $2^{\kappa^-} = \kappa$ seems to be sufficient by a more complicated proof), Rebholz established that if V = L, then R_κ holds for every successor cardinal κ. Clearly, R_κ is equivalent to the following principle if we compose each f_α with a bijection $\alpha \leftrightarrow \kappa$ for $\kappa \leq \alpha < \kappa^+$:

(R'_κ) There is a collection $\{g_\alpha | \alpha < \kappa^+\}$ of functions with $g_\alpha: \alpha \to \kappa$, so that whenever $s \in [\kappa^+]^{\kappa^-}$ and $\phi: s \to \kappa$, then $|\{\xi < \bigcap s | \forall\alpha\in s(g_\alpha(\xi) \neq \phi(\alpha))\}| < \kappa$

Also, it is easy to see, by considering $\{g_\alpha \upharpoonright \kappa \,|\alpha < \kappa^+\}$, that:

$$R_\kappa \to P_\kappa.$$

Finally, like P_κ, R_κ has a consequence in the partition calculus. To be precise, $\lambda \to [\mu:\nu]^2_\gamma$ means that whenever $f: [\lambda]^2 \to \gamma$, there is an $X \in [\lambda]^\mu$ and a $Y \in [\lambda]^\nu$ with $\bigcup X < \bigcap Y$, and $f''\{<\xi,\zeta> | \xi \in X \text{ and } \zeta \in Y\} \neq \gamma$.

Note that the negation $\lambda \not\rightarrow [\mu:\nu]^2_\gamma$ implies $\lambda \not\rightarrow [\mu+\nu]^2_\gamma$, and so provides a strong counterexample to the ordinary partition symbol. Rebholz formulated his principle with the following immediate consequence in mind:

$$R_\kappa \rightarrow \kappa^+ \not\rightarrow [\kappa:\kappa^-]^2_\kappa.$$

(If $\{g_\alpha \mid \alpha > \kappa^+\}$ is as provided by R'_κ, set $F(\beta,\alpha) = g_\alpha(\beta)$ for $\beta > \alpha$ to get a counterexample.) The conclusion here is stronger than the negative polarized partition relation entailed by P_κ, and the difference is revealing: there, each f_α need only be: $\kappa \rightarrow \kappa$, and here, we must have an elongated f_α: $\alpha \rightarrow \alpha$. Incidentally, this is the best possible limitative result, since Shelah [S] established in ZFC + GCH that: If $\kappa > \omega$ is regular and $\gamma^+ < \kappa$, then $\kappa^+ \rightarrow (\kappa+\gamma)^2_2$. Rebholz [Re] also provides an application of R_κ to the theory of free subsets for set mappings, answering a question of Máté.

The following derivation highlights the lateral approximations provided by $\vartriangleleft$ in B_κ.

<u>Theorem 5 (Kanamori)</u>: $\left(2^{\kappa^-} = \kappa \text{ and } B_\kappa\right) \rightarrow R_\kappa$.

$\vdash$ As mentioned in the proof of Theorem 4, by the Completeness condition, any $x \in T_\kappa$ has at most κ $\vartriangleleft$-predecessors. Hence, as T_κ has cardinality $> \kappa$, it must have a subset well-ordered by $\vartriangleleft$ in order-type κ^+.

So, by renaming and trimming, we might as well assume further that:

(6) $T_\kappa = \kappa^+$, and for $\alpha, \beta < \kappa^+$, we have $\alpha \lhd \beta$ <u>iff</u> $\alpha < \beta$.

To prove the theorem, it suffices to establish the more tractable R'_κ. By Theorem 4, there is a function W satisfying the clauses of Silver's Principle for our tree T. Thus, by $2^{\kappa^-} = \kappa$, for each $\delta < \kappa$ and $s \in W(\delta)$, we can enumerate ${}^{s}\kappa$ as $\{h^s_\xi \mid \delta \leq \xi < \kappa\}$, and define functions $g_\xi\colon T_\xi \to \kappa$ satisfying the condition (†), just as in the proof of Theorem 2. Finally, let us define $f_\alpha\colon \alpha \to \kappa$ for $\alpha < \kappa^+$ as follows:

If $\zeta < \alpha$, by Completeness, there is a ρ such that $\pi_\rho(\zeta) \lhd \pi_\rho(\alpha)$. Let ρ^α_ζ be the least such ρ, and set

$$f_\alpha(\zeta) = g_{\rho^\alpha_\zeta}(\pi_{\rho^\alpha_\zeta}(\alpha)).$$

This definition underscores the importance of $\lhd$; once an ordering of the top nodes is established, $\lhd$ reflects and completely approximates this ordering through the lower levels.

To verify R'_κ, let $s \in [\kappa^+]^{\kappa^-}$ and $\phi \in {}^{s}\kappa$. There is a $\delta < \kappa$ such that:

(a) $\delta \leq \xi < \kappa$ implies π_ξ"$s \in W(\xi)$, and

(b) $\delta \leq \xi < \kappa$ and $\alpha < \beta \in s$ implies $\pi_\xi(\alpha) \lhd \pi_\xi(\beta)$ and $Q(\pi_\xi(\alpha), \pi_\xi(\beta), \beta) = \alpha$.

Here, we can accomplish (b) using Completeness and Commutativity much the same as in the proof of Theorem 4. Notice the following FACT: If $\zeta < \bigcap s$, $\delta \leq \xi < \kappa$, and there is an $\alpha \in s$ such that $\pi_\xi(\zeta) \lhd \pi_\xi(\alpha)$, then for <u>every</u> $\beta \in s$ we also have $\pi_\xi(\zeta) \lhd \pi_\xi(\beta)$. This is so by transitivity of $\lhd$ if $\alpha < \beta$, and by the Coherence condition (4) if $\beta < \alpha$.

Now for some $\eta \geq \delta$, we have $h_\eta^{\pi_\delta''s} = \phi \cdot \pi_\delta^{-1}$. Set $E = \{\zeta < \bigcap s \mid \exists \alpha \in s \exists \xi < \eta \; (\pi_\xi(\zeta) \lhd \pi_\xi(\alpha)$ and $Q(\pi_\xi(\zeta), \pi_\xi(\alpha), \alpha) = \zeta)\}$. Clearly $|E| < \kappa$; E consists of the exceptional ordinals:

Suppose that $\zeta \in (\bigcap s - E)$. By Completeness and the FACT, there is a fixed ρ with $\eta \leq \rho < \kappa$ such that $\rho = \rho_\zeta^\alpha$ for every $\alpha \in s$ (where ρ_ζ^α was defined in the course of the definition of f_α). Now we can complete the proof as in Theorem 2. By condition (†) on the g_ξ's, there is an $x \in \pi_\rho''s$ such that $g_\rho(x) = h_\eta^{\pi_\delta''s}(\pi_\delta(x))$. But if $\alpha = \pi_\rho^{-1}(x) \in s$, then $f_\alpha(\zeta) = g_\rho(x) = \phi(\alpha)$. This establishes R'_κ. ⊣

The following diagram summarizes the implications in the first three sections assuming the GCH:

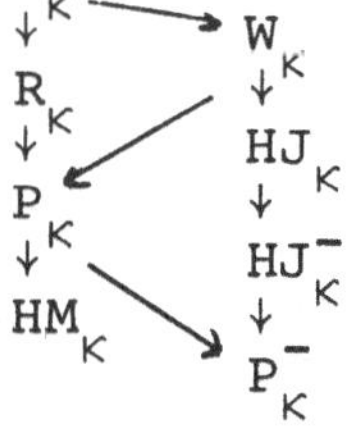

I do not know whether any converses are true.

§4. GENERALIZATIONS

This section considers versions of the various combinatorial principles also available at limit cardinals; perhaps the main interest in these generalizations lies in the consequent limitative results in the partition calculus which counterpoint the positive results available from large cardinals. So, let me provide the backdrop of historical context, first of all for the polarized partition relation.

In general, the proposition

$$(*) \qquad \begin{pmatrix} \kappa^+ \\ \kappa \end{pmatrix} \to \begin{bmatrix} \kappa \\ \kappa \end{bmatrix}_2$$

seem to hold but rarely. The earliest result along these lines was due to Erdős and Rado [ER] Theorem 48, who established (*) for $\kappa = \omega$. Hajnal [H] then established (*) for κ a measurable cardinal; see also Chudnovsky [C] and Kanamori [Ka1] for some refinements. Chudnovsky claims without proof in his paper that (*) holds for κ a weakly compact cardinal, and proofs have since been provided by Wolfsdorf [Wo], Shelah, and Kanamori [Ka2].

For successor cardinals κ, we saw in §1 that P_κ denies (*) in strong fashion. Unpublished work of

Laver [L] provides a positive consistency result: Say that a non-trivial ideal over a regular cardinal $\kappa > \omega$ is a <u>Laver ideal</u> <u>iff</u> whenever $X \subseteq P(\kappa) - I$ with $|X| \geq \kappa^+$, there is a $Y \in [X]^{\kappa^+}$ so that: whenever $Z \in [y]^{<\kappa}$, then $\bigcap Z \notin I$. Notice that any measure over a measurable cardinal κ is dual to a Laver ideal over κ. Laver noted that the existence of a Laver ideal implies (*) (where the 2 can be replaced by any ordinal $< \kappa$). Refining an argument of Kunen [Ku1], he then established the relative consistency of ω_1 carrying a Laver ideal, by forcing over a ground model satisfying ZFC and a strong large cardinal hypothesis, the existence of a huge cardinal.

The study of the even rarer

(**) $\qquad \kappa^+ \rightarrow (\alpha)^2_2$ for every $\alpha < \kappa^+$

also has a rambling history. After years of partial results and conjectures, Baumgartner and Hajnal [BH] established (**) for $\kappa = \omega$, as a consequence of a more general result which they established in elegant fashion by using Martin's Axiom and an absoluteness argument. Avoiding these tricks of the trade, Galvin [G] provided a direct proof which is a combinatorial tour de force. More recently, Tedorcevic has announced further refinements. It is not known whether (**) holds for κ a measurable cardinal; perhaps the best partial result is due to Laver. He was first to observe that if there is a Laver ideal over κ, then $\kappa^+ \to (\kappa+\kappa+1,\alpha)^2_2$ for every $\alpha < \kappa^+$. I established Laver's result without knowing of it, and a full proof is provided in [Ka4]. Also in [Ka4] is the result that if κ is a weakly compact cardinal, then $\kappa^+ \to (\kappa+\kappa+1,[\kappa:\kappa^+])^2_2$, a technical statement somewhat stronger than $\kappa^+ \to [\kappa:\kappa]^2_2$, which already follows from the known relation (*) for weakly compact cardinals. For successor cardinals, we saw in §3 that R_κ denies (**) in strong fashion. In the positive direction, there is again Laver's result about a Laver ideal over ω_1, starting with the consistency strength of a huge cardinal. Gray also has some partial positive results.

Turning to the subject at hand, just as P_κ and R_κ deny partition relations for successor cardinals, there are weaker versions which delimit the situation for possibly limit cardinals.

(wP_κ) There is a collection $\{f_\alpha \mid \alpha < \kappa^+\} \subseteq {}^\kappa\kappa$ so that whenever $s \in [\kappa^+]^\kappa$ and $\phi \in {}^s\kappa$, we have $|\{\xi < \kappa \mid \forall\alpha\in s(f_\alpha(\xi) \neq \phi(\alpha))\}| < \kappa$.

(wR_κ) There is a collection $\{f_\alpha \mid \alpha < \kappa^+\}$ of functions $f_\alpha\colon \alpha \to \alpha$ so that whenever $s \in [\kappa^+]^\kappa$ and ϕ is a regressive function with domain = s, then $|\{\xi < \bigcap s \mid \forall\alpha\in s(f_\alpha(\xi) \neq \phi(\alpha))\}| < \kappa$.

There are versions of the other combinatorial principles with the requisite strength (one is stated for W_κ in [Ka3]), but to discuss them would take us somewhat afield. In direct analogy to previous results, we have:

$$wP_\kappa \quad \rightarrow \quad \begin{pmatrix}\kappa^+\\ \kappa\end{pmatrix} \not\rightarrow \begin{bmatrix}\kappa\\ \kappa\end{bmatrix}_2$$

$$wR_\kappa \quad \rightarrow \quad \kappa^+ \not\rightarrow [\kappa:\kappa]^2_\kappa$$

I established [Ka2][Ka4] the consistency of these propositions via a forcing which does not (too much) disturb the universe; e.g. for the stronger wR_κ,

<u>Theorem 6 (Kanamori)</u>: If the ground model satisfies $\kappa^{<\kappa} = \kappa$, then there is a κ^+-c.c., $<\kappa$-distributive forcing extension in which wR_κ holds. (Furthermore, properties like the Mahloness of κ are preserved.)

The proof involves a new and elegant kind of density argument, first seen in the work of Shelah. By itself, this is a piecemeal result, and to genuinely contrast the positive partition relations from large cardinals, the actual situation in L must be ascertained. Recent and continuing work of Shelah and Stanley [SS1][SS2] and Velleman [V] have made the formidable apparatus of the $(\kappa,1)$-morass (a gap-1 morass at κ) more tractable (at least for some) by providing a Martin's Axiom-type characterization. That is, certain partial orders and collections of dense sets are described, and the existence of a morass is shown to be equivalent to the proposition that for every such partial order and every such collection F of dense sets, there is an F-generic filter in the usual sense. The partial order used in the proof of Theorem 6 is a paradigm case of a canonical limit partial order, in the sense of Shelah and Stanley. It was to handle such orders that led Shelah and Stanley to extend their characterization of morasses. They show how canonical limit partial orders can be accommodated in a Martin's Axiom-type characterization for $(\kappa,1)$-morasses "with built-in $\Diamond$ principle", when there is a non-reflecting stationary subset of κ, i.e. an $S \subseteq \kappa$ which is stationary in κ yet $S \cap \alpha$ is not stationary in α for any $\alpha < \kappa$. They establish that such morasses with built-in $\Diamond$ principle exist in L, and, of course, it is a well-known result of Jensen [Je] that in L, a regular $\kappa > \omega$ is not weakly compact iff there is a non-reflecting stationary subset of κ. Velleman is also developing a scheme along similar lines, but with a more concise formulation. Assuming that the partial order used in Theorem 6 fits into either the Shelah-Stanley or Velleman scheme in its final form, we have the following characterization of weak compactness in L:

Theorem 7: If V = L, then the following are equivalent for regular $\kappa > \omega$:

(i) κ is not weakly compact

(ii) wR_κ

(iii) wP_κ

(iv) $\begin{pmatrix}\kappa^+\\ \kappa\end{pmatrix} \not\to \begin{bmatrix}\kappa\\ \kappa\end{bmatrix}_\kappa$

(v) $\kappa^+ \not\to [\kappa:\kappa]^2_\kappa$

This is the heralded counterpoint to large cardinals.

REFERENCES

[BSV] B. Balcar, P. Simon & P. Vojtáš, Refinement properties and extending of filters, to appear.

[BH] J. Baumgartner & A. Hajnal, A proof (involving Martin's Axiom) of a partition relation, Fund. Math. 78(1973), 193-203.

[Bu1] J. Burgess, Forcing, in: J. Barwise, ed., Handbook of Mathematical Logic, North Holland (Amsterdam 1977), 403-452.

[Bu2] J. Burgess, On a set-mapping problem of Hajnal and Máté, Acta Sci. Math. 41(1979), 283-288.

[C] G. Chudnovsky, Combinatorial properties of compact cardinals, in: Infinite and Finite Sets, Colloquia Mathematica Societatis Janos Bolyai 10, North Holland (Amsterdam 1975), 289-306.

[D] K. Devlin, On hereditarily separable Hausdorff spaces in the constructible universe, Fund. Math. 82(1974), 1-10.

[EHR] P. Erdős, A. Hajnal & R. Rado, Partition relations for cardinal numbers, Acta Math. Acad. Sci. Hungar. 16(1965), 93-196.

[ER] P. Erdős & R. Rado, A partition calculus in set theory, Bull. Amer. Math. Soc. 62(1956), 427-489.

[G] F. Galvin, On a partition theorem of Baumgartner and Hajnal, in: Infinite and Finite Sets, Colloquia Mathematica Societatis Janos Bolyai 10, North Holland (Amsterdam 1975) 711-729.

[GH] F. Galvin & A. Hajnal, Inequalities for cardinal powers, Annals Math. 101(1975), 491-498.

[H] A. Hajnal, On some combinatorial problems involving large cardinals, Fund. Math. 69(1970), 39-53.

[HJ1] A. Hajnal & I. Juhász, A consistency result concerning hereditarily α-separable spaces, Akademie van Wetenschappen Amsterdam Proceedings A76 (or Indag. Math. 35)(1973), 301-307.

[HJ2] A. Hajnal & I. Juhász, A consistency result concerning hereditarily α-Lindelöf spaces, Acta Math. Acad. Sci. Hungar. 24(1973), 307-312.

[HM] A. Hajnal & A. Máté, Set mappings, partitions, and chromatic numbers, in: H. Rose & J. Sheperdson, eds., Logic Colloquium '73, North Holland (Amsterdam 1975).

[Je] R. Jensen, The fine structure of the constructible hierarchy, Annals Math. Logic 4(1972), 229-308.

[Ju] I. Juhász, Consistency results in topology, in: J. Barwise, ed., Handbook of Mathematical Logic, North Holland (Amsterdam 1977), 503-522.

[Ka1] A. Kanamori, Some combinatorics involving ultrafilters, Fund. Math. 100(1978), 145-155.

[Ka2] A. Kanamori, Morass-level combinatorial principles, to appear in the proceedings of the Patras 1980 conference.

[Ka3] A. Kanamori, On Silver's and related principles, to appear in the proceedings of the Prague 1980 conference.

[Ka4] A. Kanamori, Partitions of κ^+, to appear.

[Ku1] K. Kunen, Saturated ideals, Jour. Sym. Logic 43(1978), 65-76.

[Ku2] K. Kunen, Strong S and L spaces under MA, in: G. Reed, ed., Set-Theoretic Topology, Academic Press (New York 1977), 265-268.

[L] R. Laver, Strong saturation properties of ideals (abstract), Jour. Sym. Logic 43(1978), 371.

[P] K. Prikry, On a problem of Erdős, Hajnal and Rado, Discrete Math. 2(1972), 51-59.

[Re] J. Rebholz, Consequences of morass and diamond, Annals Math. Logic 7(1974), 361-385.

[Ru] M.E. Rudin, Lectures in Set Theoretic Topology, Regional Conference Series in Mathematics 23, A.M.S. (Providence RI 1975).

[S] S. Shelah, Notes on combinatorial set theory, Israel Jour. Math. 14(1973), 262-277.

[SS1] S. Shelah & L. Stanley, S-Forcing, I: a "black box" theorem for morasses, with applications: super-Souslin trees and generalized Martin's Axiom, to appear in Israel Jour. Math.

[SS2] S. Shelah & L. Stanley, S-Forcing, II, to appear.

[Sz] A. Szymański, Some Baire Category type theorems for $U(\omega_1)$, Commentationes Mathematicae Universitatis Carolinae 20(3)(1979), 519-528.

[V] D. Velleman, Doctoral dissertation, University of Wisconsin--Madison, 1980. To appear as a Springer Lecture Notes in Mathematics.

[Wi] N. Williams, Combinatorial Set Theory, North Holland (Amsterdam 1977).

[Wo] K. Wolfsdorf, Der beweis eines satzes von G. Chudnovsky, Archiv Math. Logik 20(1980), 161-171.

A SHORT COURSE ON GAP-ONE MORASSES WITH A REVIEW OF THE FINE STRUCTURE OF L.

Lee Stanley

Maths. Pures - Les Cézeaux
Université de Clermont II
63170 Aubière, France.

§1. Introduction

This paper is based on a series of lectures given at the Cambridge Summer School in Set Theory in 1978. Those lectures were primarily devoted to gap-one morasses, but it seemed reasonable in the context of this paper to review the fine structure of L and collect together the various remarks and modifications circulating in the folklore which together constitute a serious rationalization of this theory as it first appeared in [6] or [2]. I should emphasize that there is little or nothing here which is not already implicit in Jensen's original treatment, and that most of the cognoscenti have probably worked out similar improvements in private. Finally, it's worth noting that the approach taken here becomes practically indispensable when dealing with models other than L; in fact, Jensen and Dodd chose exactly this approach in [5]; my exposition owes much to their treatment.

My approach to morasses owes much to Jensen, of course, but also to Devlin, Silver and Burgess. A notable feature is the early introduction of a less complex structure, the coarse morass, which in a very real sense embodies the condensation arguments needed to prove combinatorial principles like $\diamondsuit$ and

$\diamondsuit^+$ and which, like these principles, does not need the fine structure theory. As the title indicates, I shall only be concrened with gap-one morasses, thus, unless otherwise indicated "morass" means "gap-one morass". This material has also appeared in [2], and is introduced rather sketchily in Chapter 1 of [9]. The proof outlined in [9] that morasses exist in L will be fully developed in this paper. It avoids the unnecessary computations involved in proving ($\heartsuit$) (cf. [2]), it brings out more clearly the way the morass properties arise as abstractions from the situation in L and thus as abstractions from the fine structure theory of L. This proof has the added advantage of generalizing relatively straightforwardly to higher-gap morasses.

Jensen had a proof along these lines no later than 1972; the proof given was worked out in the course of an informal seminar on morasses in Berkeley in early 1972. The basis "map-splitting" lemma, (2.11), attributed to Burgess in [9] should probably be jointly attributed to Jensen and Burgess; the proof is new, more in the spirit of certain arguments of [4]. More recently, Silver and Richardson have developed an alternative construction of morasses in L via Silver's machines.

The approach taken to adding morasses by forcing also was developed by Jensen and antedates 1972. It is essentially that presented in [9], in that the generic object is a structure simpler than a morass (Jensen called it a premorass) but which can be thinned in the generic extension to yield a morass. In the meantime a much simpler set of forcing conditions has been worked out by D. Velleman [10] who has kindly permitted me to present them here. It seems unlikely however that there is a generalization of Velleman-style forcing for higher gap morasses.

I work in ZFC, except in §3 where, when working in L, I temporarily assume V = L for convenience. Notation, terminology etc. which is not standard is either intended to have a clear meaning (e.g. card (x) for cardinal of x, o.t.(x) for order type of x), is to be found in [6], or [2], or is introduced explicitly as needed. I suppose the reader is familiar at least with §§1,2 of [6] (see the discussion of the beginning of §2). §2 is devoted to the fine structure theory of L; §3 to morasses and the proof that they exist in L. In §4 I give some combinatorial applications of morasses; for others, see [1],[7] and the forthcoming [8], [10] where it is proved that the existence of morasses is equivalent to certain Martin's Axiom-type principles, thus providing a new and very systematic way of applying morasses which is exploited in [8]. In §5 I show how morasses can be added by forcing.

§2 THE FINE STRUCTURE OF L, NOW THAT THE DUST HAS CLEARED.

My goal in this section is to rework the theory of projecta and standard codes as presented in §§3,4 of [6]. My point of departure is §§1,2 of [6] which remains an excellent source for the basic properties of rudimentary functions, the J_α hierarchy and the auxiliary S_α hierarchy. Since this paper does not pretend to be self-contained, I shall assume knowledge of this material. Nevertheless, before getting under way it seems reasonable to point out a number of minor improvements and/or corrections which have been made in the meantime, and to summarize some particularly important points.

First, the J_α's are not only transitive, but closed for transitive closure; this is proved in a more general setting in [5]. Second, by slightly modifying the finite basis for the rudimentary functions which gives the definition of the S_α's, the S_α's may actually be taken to be transitive. Third, for a correct account of the exact relationship between the J_α-hierarchy and the L_α-hierarchy the reader should consult [3]; it will suffice here to recall that the subsets of J_α in $J_{\alpha+1}$ are just those which are parameter-definable over J_α, and that if $\alpha = \omega.\alpha$ then $J_\alpha = L_\alpha$. Finally, I'll recall a number of facts which figure prominently in all that follows.

a) For all α, all $\beta < \alpha$, all $n < \omega$, $(S_\eta : \eta < \omega.\beta+n)$, $(J_\gamma : \gamma < \beta)$, $(<_J|S_\eta : \eta < \omega.\beta+n) \in J_\alpha$.

b) There are Σ_0 formulas $\psi(f,\alpha)$, $\psi'(f,\alpha)$ which hold, respectively, iff ($\alpha \in OR$ and $f = (S_\eta : \eta < \alpha)$) and iff ($\alpha \in OR$ and $f = (<_J|S_\eta : \eta < \alpha)$.

c) For all α, $(S_\eta : \eta < \omega.\alpha)$, $(J_\beta : \beta < \alpha)$, $(<_J|S_\eta : \eta < \omega.\alpha)$, $<_J|J_\alpha$ are uniformly $\Sigma_1(J_\alpha)$.

d) There is a Π_2 sentence θ such that for transitive rudimentarily closed M, $M \models \theta$ iff for some α, $M = J_\alpha$. θ can be taken to be e.g.: $\forall\beta\ \exists f\psi(f,\beta)\wedge\forall x\ \exists f\ \exists\gamma>\beta$ $[\psi(f,\beta)\wedge x\in f(\gamma)]$. In what follows I'll designate θ by "I am a J_α".

e) The Σ_1 satisfaction relation $\models^1_{M,A}$ is uniformly $\Sigma_1(M,A)$ for transitive rudimentarily-closed amenable structures (M,A); further, uniformly Σ_1 uniformisation holds for such structures, when M is a J_α, and $h_{\alpha,A}$, the canonical Σ_1-Skolen function for (J_α,A) is the uniform Σ_1-uniformisation of $\models^1_{J_\alpha,A}$. When $A = \emptyset$, I suppress mention of it. $h_{\alpha,A}$ yields minimal Σ_1-elementary substructures of (J_α,A) when applied to sets of the form $\omega\times B\times\{x\}$, whenever $B\cup\{x\} \subseteq J_\alpha$ and B is closed for some reasonable pairing function which, together with its coordinate-wise inverses is $\Sigma_1(J_\alpha,A)$ definable in parameter x. The range of $h_{\alpha,A}$ on $\omega\times B\times\{x\}$ is just the set of denotations of Σ_1-terms of (J_α,A) in parameters from $B\cup\{x\}$. For non-amenable (J_α,A) the Σ_1-Skolem function can also be defined with all these properties except the Σ_1-definability over (J_α,A); $h_{\alpha,A}$ will also denote the Σ_1-Skolem function in this setting. The Condensation Lemma states that if $X \prec_1 J_\alpha$ then for some

$\beta, X \cong J_\beta$. For all α, there is a finite set of ordinals $a \in J_\alpha$ such that $J_\alpha = h''_\alpha(\omega\times\omega.\alpha\times\{a\})$. If π is a collapsing map then for $x \in \operatorname{dom} \pi\cap L$ $\pi(x) \leq_L x$, $\pi(x) \leq_J x$.

I can now get started on the fine structure theory of L.

(2.1) <u>DEFINITION</u>: Let (J_α, A) be amenable. The Σ_1-projectum of α and A, $\rho^1_{\alpha,A}$, is the least $\rho \leq \alpha$ such that:

(*): There is $a \subseteq J_\alpha$ such that $a\cap J_\rho \notin J_\alpha$ and which is $\Sigma_1(J_\alpha, A)$ parameter definable.

$p^1_{\alpha,A}$, the Σ_1-standard parameter for α, A is the $<_J$-least $p \in J_\alpha$ such that some a as in (*) is $\Sigma_1(J_\alpha, A)$ in p. $A^1_{\alpha,A}$, the Σ_1-standard code for α, A, is $\models^1_{(J_\alpha, A, p)} \cap J_\rho$, where $p = p^1_{\alpha,A}$; that is, $(i, \vec{x}) \in A^1_{\alpha,A}$ iff $(J_\alpha, A) \models \phi_i[\vec{x}, p]$ and $\vec{x} \in J_\rho$, where $(\phi_i: i < \omega)$ is some reasonable fixed Godël-numbering of Σ_1-formulas. (J_α, A) is 1-sound iff $J_\alpha = h_{\alpha,A}''(\omega\times\omega.\rho\times\{p\})$, i.e. iff all elements of J_α are Σ_1-definable in parameters from $\omega.\rho\cup\{p\}$. In keeping with the above convention (e), I write $\rho^1_\alpha, p^1_\alpha, A^1_\alpha$, "$J_\alpha$ is 1-sound" if $A = \emptyset$.

(2.2) Here is an example of amenable (J_α, A) which is not 1-sound. Let $x \in \mathcal{P}(\omega)\setminus L$ be such that $\aleph_1^{L[x]} = \aleph_1^L$, and suppose $V = L[x]$. Let $\alpha = \omega_1+1$, $A = \{\omega_1+n: n \in x\}$. Clearly (J_α, A) is amenable, x is $\Sigma_1(J_\alpha, A)$ in parameter ω_1. Hence $\rho^1_{\alpha,A} = 1$ and so (J_α, A) is not 1-sound.

In what follows, I do not always give the most compact proof possible if I feel that it would mask "what is going on". On the other hand, in order to develop the theory as harmoniously as possible, I've had no qualms about making heavy use of the Condensation Lemma even when it's not strictly necessary. This has the drawback, for the reader interested in models other than L, that it is not always exactly clear what generalizes and what does not, but forewarned is forearmed.

(2.3) PROPOSITION: a) If $\eta \leq \alpha$ and $J_\alpha = h_{\alpha,A}$"$(\omega\times\omega.\eta\times\{x\})$ where $x \in J_\alpha$, then there is $a \subseteq J_\alpha$ such that $a\cap\omega.\eta \notin J_\alpha$, which is $\Sigma_1(J_\alpha,A)$ in x; hence $\eta \geq p^1_{\alpha,A}$.

b) If $\rho^1_{\alpha,A} < \alpha$ then $J_\alpha \models \omega.\rho^1_{\alpha,A}$ is a cardinal.

c) If $\eta < \omega.\rho^1_{\alpha,A}$, $x \subseteq S_\eta$ then $x \in J_\alpha$ iff $x \in J_\rho$ (where $\rho = \rho^1_{\alpha,A}$); hence $(J_\rho, A^1_{\alpha,A})$ is amenable, and more generally if $x \subseteq S_\eta$, $\eta < \omega.\rho$, and x is $\Sigma_1(J_\alpha,A)$ parameter-definable then $x \in J_\rho$.

Proof: a) (this is essentially the argument of example (2) of p.256 of [6]). First, define, for $\zeta < \omega.\alpha$, (ζ) = the largest limit ordinal $\leq \zeta$, and let $[\zeta] = \zeta-(\zeta)$. If $[\zeta] = 0$, set $(\zeta)_0 = 0$, $(\zeta)_1 = (\zeta)$; otherwise, let i,j be such that $[\zeta] = 2^i.(2j+1)$ and set $(\zeta)_0 = i$, $(\zeta)_1 = (\zeta)+j$. Then $g(\zeta) = ((\zeta)_0,(\zeta)_1)$, is $\Sigma_1(J_\alpha,A)$ (in fact, $\Sigma_1(J_\alpha)$, in fact $\Delta_1(J_\alpha)$), and for $\gamma \leq \alpha$, $g: \omega.\gamma \to \omega\times\omega.\gamma$ is a bijection.

Set $\zeta \in a$ iff $\zeta \in OR$, $h_{\alpha,A}((\zeta)_0, (\zeta)_1, x)$ is defined and $\zeta \notin h_{\alpha,A}((\zeta)_0,(\zeta)_1,x)$. Clearly a is as required. The second assertion is clear from the first.

b) Suppose $\rho = \rho_{\alpha,A}$, $\eta < \rho$, $f: \omega.\eta \to \omega.\rho$, $f \in J_\alpha$. Let $a \in P(\omega.\rho)\backslash J_\alpha$ be $\Sigma_1(J_\alpha,A)$ in $p^1_{\alpha,A}$. Then, since $\eta < \rho$, $f^{-1}[a] \in J_\alpha$, because $f^{-1}[a]$ is $\Sigma_1(J_\alpha,A)$ in $(p^1_{\alpha,A},f)$. Since $f \in J_\alpha$, f"$f^{-1}[a] \in J_\alpha$, so, if f were onto f"$f^{-1}[a]$ would be a, and would lie in J_α, which is absurd.

c) (WARNING: CONDENSATION LEMMA) Let $x \subseteq S_\eta$ and let $\omega.\gamma$ be the largest limit ordinal $\leq \eta$. Let $Y = h_\alpha$"$(\omega\times\omega.\gamma\times\{x\})$. Thus $S_\eta \subseteq Y \prec_1 J_\alpha$, and so setting $\pi: J_{\bar\alpha} \xrightarrow{\sim} Y$, $\pi^{-1}(x) = x$. Since $J_{\bar\alpha} = h_{\bar\alpha}$"$(\omega\times\omega.\gamma\times\{x\})$, by a) $\bar\alpha < \omega.\rho$: i.e. $x \in J_\rho$. The remaining assertions are now clear.

(2.4) PROPOSITION: If $1 \leq n < \omega$ and $x \subset J_\rho$ is $\underset{\sim}{\Sigma}_n(J_\rho,A^1_{\alpha,A})$ then x is $\underset{\sim}{\Sigma}_{n+1}(J_\alpha,A)$ $(\rho = \rho^1_{\alpha,A})$.

Proof: (this is Lemma 3.2 of [6]; the proof is the same as there) Let $B = A^1_{\alpha,A}$. It clearly suffices to show that if $x \subset J_\rho$ is $\underset{\sim}{\Sigma}_0(J_\rho,B)$ then x is $\underset{\sim}{\Sigma}_2(J_\alpha,A)$. For this it suffices to show that $b(u) = B\cap u$ is $\underset{\sim}{\Sigma}_2(J_\alpha,A)$. There are two cases.

CASE 1: There are $\gamma < \omega\rho$, $g: \gamma \to_{\text{cofinal}} \omega\alpha$ such that g is $\Sigma_1(J_\alpha,A)$. In this case Jensen's trick is first to replace B by a B' which is $\underset{\sim}{\Delta}_1(J_\alpha,A)$ and such that B is $\Sigma_0(J_\rho,B')$. Thus, if x is $\underset{\sim}{\Sigma}_1(J_\rho,B)$, x is $\underset{\sim}{\Sigma}_1(J_\rho,B')$, so it suffices to prove that $b'(u) = B'\cap u$ is $\underset{\sim}{\Sigma}_2(J_\alpha,A)$. This is easy since $y = b'(u) \iff (\forall z)(z\epsilon y \iff z\epsilon u \wedge z\epsilon B')$. The matrix is $\underset{\sim}{\Delta}_1(J_\alpha,A)$ and so $y = b'(u)$ is in fact $\Pi_1(J_\alpha,A)$. It remains to find B' with the desired properties. Set $(\nu,y) \in B'$ iff $\nu < \gamma$ and $(\exists z \in S_{g(\nu)})((z,y) \in \bar{B})$, where $\bar{B}$ is $\underset{\sim}{\Sigma}_0(J_\alpha,A)$ such that $B = \{y\epsilon J_\rho: (\exists z\epsilon J_\alpha)(z,y)\epsilon\bar{B})\}$. B' is $\underset{\sim}{\Delta}_1(J_\alpha,A)$ because the definition can be written out as: $\nu < \gamma$ and

$$(\exists\xi)(\exists w)(\text{"}\xi = g(\nu)\text{"}\wedge\text{"}w = S_\xi\text{"}\wedge(\exists z\epsilon w)((z,y)\epsilon\bar{B})),\text{ or } \nu < \gamma$$

and

$$(\forall\xi)(\forall w)((\text{"}\xi = g(\nu)\text{"}\wedge\text{"}w = S_\xi\text{"}) \Longrightarrow (\exists z\epsilon w)((z,y)\epsilon\bar{B})).$$

Finally, $B = \{y: (\exists\nu < \gamma)((\nu,y) \in B')\}$.

CASE 2: There are no such γ,g. In this case we have a replacement-like property for $\underset{\sim}{\Sigma}_1(J_\alpha,A)$ relations and elements of J_ρ:

If Rzy is $\underset{\sim}{\Sigma}_1(J_\alpha,A)$ then so is $R'wy$ where $R'wy \iff w \in J_\rho$ and $(\forall z\epsilon w)Rzy$. This is because, roughly speaking, the hypothesis (that there are no such γ,g) guarantees that there is a bound $\nu < \omega\alpha$ such that if Rzy is $\exists u\, \bar{R}uzy$ where $\bar{R}$ is $\underset{\sim}{\Sigma}_0(J_\alpha,A)$, then for all $z \in w$, $\exists u\, \bar{R}uzy \iff (\exists u \in S_\nu)\bar{R}uzy$. The Σ_1-definition of R' comes from existentially quantifying over such bounds ν and bounding the existential quantifier on u by S_ν. But then

$$y = b(u) \leftrightarrow \overset{1}{(\forall x \in y)(x \in u \wedge x \in B)} \wedge \overset{2}{(\forall x \in u)(x \in B \rightarrow x \in y)}$$

Now 2 is $\underset{\sim}{\Pi}_1$ and, by the preceding, 1 is $\underset{\sim}{\Sigma}_1(J_\alpha,A)$. A Boolean combination of $\underset{\sim}{\Pi}_1$ and $\underset{\sim}{\Sigma}_1$ is $\underset{\sim}{\Sigma}_2(J_\alpha,A)$. Note that this proof works for any B which is $\underset{\sim}{\Sigma}_1(J_\alpha,A)$.

(2.5) <u>COROLLARY:</u> If (J_α,A) is 1-sound then for all $x \subseteq J_\rho$, all positive $n < \omega$, x is $\underset{\sim}{\Sigma}_n(J_\rho,B)$ iff x is $\underset{\sim}{\Sigma}_{n+1}(J_\alpha,A)$ $(\rho = \rho^1_{\alpha,A},\ B = A^1_{\alpha,A})$.

<u>Proof:</u> By (2.4), and by easy arguments as implicit in the proof of (2.4), it suffices to prove that if $x \subseteq J_\rho$ is $\underset{\sim}{\Sigma}_1(J_\alpha,A)$, then x is $\underset{\sim}{\Sigma}_0(J_\rho,B)$. So, suppose $\vec{y} = (y_0,\ldots,y_{m-1}) \in J_\alpha$, ϕ is Σ_1 such that $x = \{z \in J_\rho : (J_\alpha,A) \models \phi[z,\vec{y}]\}$. By soundness, for $j < m$ there are $i_j < \omega$, $\xi_j < \omega.\rho$ such that $y_j = h_{\alpha,A}(i_j,\xi_j,p)$ $(p = p^1_{\alpha,A})$. Accordingly, letting θ be the Σ_1 formula in parameters i_j,ξ_j, $(j<m)$, p, $\theta = \exists y_0 \ldots y_{n-1}(\bigwedge_{j<m} "y_j = h_{\alpha,A}(i_j,\xi_j,p)" \wedge \phi(z,y_0,\ldots,y_{n-1}))$, $x = \{z \in J_\rho : (J_\alpha,A) \models \theta(z,\vec{i},\vec{\xi},p)\}$. Letting k be such that θ is the kth Σ_1-formula, $x = \{z: (k,z,\vec{i},\vec{\xi}) \in B\}$. Thus, x is $\Sigma_0(J_\rho,B)$ in parameters $k,\vec{i},\vec{\xi}$.

(2.6) <u>PROPOSITION:</u> Let (J_ρ,B) be amenable. There is at most one (J_α,A) which is amenable, 1-sound and such that $\rho = \rho^1_{\alpha,A}$, $B = A^1_{\alpha,A}$.

<u>Proof:</u> Suppose that $(J_\alpha,A),(J_{\alpha'},A')$ both have these properties. Let $\theta(x,i,y,z)$ express "$x = h_{\eta,E}(i,y,z)$" when interpreted in amenable rudimentarily closed (J_η,E). Let j be such that the jth Σ_1-formula is $\exists x\theta(x,i,y,z)$. Let $\tilde{M} = \{(i,y):(j,i,y) \in B\}$. Let k be such that the kth Σ_1-formula is $\exists x\exists x'(\theta(x,i,y,z) \wedge \theta(x',i',y',z) \wedge x = x')$. Let $R = \{((i,y),(i',y')):(k,i,y,i',y') \in B\}$. Let ℓ be such that the ℓth Σ_1-formula is $\exists x\exists x'(\theta(x,i,y,z)\wedge\theta(x',i',y',z)\wedge x \in x')$. Let $E = \{((i,y),(i',y')):(\ell,i,y,i',y') \in B\}$. Let m be such that the

$m^{\underline{th}}$ Σ_1-formula is $\exists x(\theta(x,i,y,z) \wedge \underline{U}(x))$ (here $\underline{U}$ is the additional unary predicate symbol interpreted by A,A' respectively). Let $U = \{(i,y):(m,i,y) \in B\}$.

Then, by the hypotheses, R is an equivalence relation on $\tilde{M}$ and is a congruence for E and U; E is a well-founded, extensional-mod-R relation on $\tilde{M}$ and $(J_\alpha,\in,A) \tilde{=} (\tilde{M}/R,E,U) \tilde{=} (J_{\alpha'},\in,A')$, the isomorphisms being $(i,y) \longmapsto h_{\alpha,A}(i,y,p)$, and $(i,y) \longmapsto h_{\alpha',A}(i,y,p')$ $(p = p^1_{\alpha,A}, p' = p^1_{\alpha',A'})$. Thus $\alpha = \alpha'$, $A = A'$.

(2.7) <u>COROLLARY:</u> Let $(J_\alpha,A),(J_{\bar{\alpha}},\bar{A})$ be amenable and 1-sound. Let $p = p^1_{\alpha,A}$, $\bar{p} = p^1_{\bar{\alpha},\bar{A}}$, $\rho = \rho^1_{\alpha,A}$, $\bar{\rho} = \rho^1_{\bar{\alpha},\bar{A}}$, $B = A^1_{\alpha,A}$, $\bar{B} = A^1_{\bar{\alpha},\bar{A}}$. Suppose $m < \omega$, $f:(J_{\bar{\alpha}},\bar{A}) \rightarrow_1 (J_\alpha,A)$, $f(\bar{p}) = p$ and $f|J_{\bar{\rho}}: (J_{\bar{\rho}},\bar{B}) \rightarrow_m (J_\rho,B)$. Then $f: (J_{\bar{\alpha}},\bar{A}) \rightarrow_{m+1} (J_\alpha,A)$.

<u>Proof</u>: If $m = 0$, there is nothing to prove, so suppose $m > 0$. Let $\mathcal{M} = (\tilde{M}/R,E,U)$ be the model constructed in (2.6) for (J_α,A), and let $\bar{\mathcal{M}}$ be the analogous model constructed for $(J_{\bar{\alpha}},\bar{A})$. Then $\models^1_{\mathcal{M}}$, $\models^1_{\bar{\mathcal{M}}}$ are $\Sigma_0(J_\rho,B)$, $\Sigma_0(J_{\bar{\rho}},\bar{B})$ by the same Σ_0-formula. Accordingly, for each $n < \omega$ $\models^{n+1}_{\mathcal{M}}$, $\models^{n+1}_{\bar{\mathcal{M}}}$ are $\Sigma_n(J_\rho,B)$, $\Sigma_n(J_{\bar{\rho}},\bar{B})$ respectively by the same Σ_n formula. As $\mathcal{M} \tilde{=} (J_\alpha,A)$, $\bar{\mathcal{M}} \tilde{=} (J_{\bar{\alpha}},\bar{A})$ and the isomorphisms are $\Sigma_1(J_\alpha,A)$ in p, $\Sigma_1(J_{\bar{\alpha}},\bar{A})$ in $\bar{p}$ by the same Σ_1-formula, and since $f: (J_{\bar{\alpha}},\bar{A}) \rightarrow_1 (J_\alpha,A)$, it is easy to see that $f: (J_{\bar{\alpha}},\bar{A}) \rightarrow_{m+1} (J_\alpha,A)$.

(2.8) <u>DEFINITION:</u> Set $\rho^0_\beta = \beta$, $p^0_\beta = A^0_\beta = \emptyset$ and say that J_β is o-sound for all β. For $n < \omega$, having defined ρ^n_β, p^n_β, A^n_β and the notion of n-soundness of β, set $\alpha = \rho_\beta$, $A = A^n_\beta$, and define $\rho^{n+1}_\beta = \rho^1_{\alpha,A}$, $p^{n+1}_\beta = p^1_{\alpha,A}$, $A_\beta = A^1_{\alpha,A}$, and say that β is n+1-sound iff β is n-sound and (J_α,A) is 1-sound. Note that this is coherent definition since, inductively using (2.3), (J_α,A) is amenable and that for n = o the above definitions coincide with those of (2.1).

The next lemma is the basic fine structure lemma, and the Condensation Lemma plays a central role. The proof uses (2.3)-(2.7). The essential construction is that of STEPS 1,2 below, the rest being essentially checking details.

(2.9) LEMMA: For all β, for all $n < \omega$, setting $\alpha = \rho^n_\beta$, $A = A^n_\beta$:

a) For $1 \leq m < \omega$ and $x \subseteq J_\alpha$, x is $\underset{\sim}{\Sigma}_m(J_\alpha, A)$ iff x is $\underset{\sim}{\Sigma}_{m+n}(J_\beta)$; if for some $\eta < \omega.\alpha$, $x \subseteq S_\eta$ then $x \in J_\alpha$ iff $x \in J_\beta$.

b) α is the least ordinal $\alpha' \leq \beta$ such that there is $a \in P(\omega\alpha')\backslash J_\beta$ which is $\underset{\sim}{\Sigma}_n(J_\beta)$, p^n_β is the $<_J$-least $p \in J_\beta$ for which there is such an a which is $\Sigma_n(J_\beta)$ in p. Further, there is a map g from a subset of $\omega\alpha$ onto J_β which is $\underset{\sim}{\Sigma}_1(J_\beta)$ if $n = o$ and is $\underset{\sim}{\Sigma}_n(J_\beta)$ if $n > o$; α is minimal for the existence of such a g, when $n > 0$.

c) J_β is $n+1$-sound.

d) ("Downward Extension of Embeddings"): Let $\rho = \rho^{n+1}_\beta$, $B = A^{n+1}_\beta$. Suppose that $m < \omega$, that $(J_{\bar\rho}, \bar B)$ is amenable, that $f: (J_{\bar\rho}, \bar B) \to_m (J_\rho, B)$. Then there are unique $\bar\beta, \tilde f$ such that: i) $\tilde f \supseteq f$, $\bar\rho = \rho^{n+1}_{\bar\beta}$, $\bar B = A^{n+1}_{\bar\beta}$, $\bar\beta$ is $n+1$-sound, ii) for $k \leq n+1$, let $k' = (n+1)-k$, let $\rho_k = \rho^{k'}_\beta$, $\bar\rho_k = \rho^{k'}_{\bar\beta}$, $A_k = A^{k'}_\beta$, $\bar A_k = A^{k'}_{\bar\beta}$, $p_k = p^{k'}_\beta$, $\bar p_k = p^{k'}_{\bar\beta}$, [thus $\rho_{n+1} = \beta, A_{n+1} = \emptyset$, $\bar\rho_{n+1} = \bar\beta$, $\bar A_{n+1} = \emptyset$, $\rho_o = \rho$, $\bar\rho_o = \bar\rho$, $A_o = B$, $\bar A_o = \bar B$]. THEN: for $k \leq n+1$, $\tilde f(\bar p_k) = p_k$, $\tilde f | J_{\bar\rho_k} : (J_{\bar\rho_k}, \bar A_k) \to_{m+k} (J_{\rho_k}, A_k)$.

Proof: The proof is by induction on n for all β at once; a), b) are trivial for n = o. STEP 1 is to prove c) for n assuming a),

b) for $\bar{n} \le n$ and c), d) for $\bar{n} < n$. STEP 2 is to prove d) for n assuming a),b),c) for $\bar{n} \le n$, and d) for $\bar{n} < n$. STEP 3 is to prove a) for n+1 assuming a)-d) for $\bar{n} \le n$ and STEP 4 is to prove b) for n+1 assuming b),c),d) for $\bar{n} \le n+1$ and a) for $\bar{n} \le n$.

STEP 1: By hypothesis if n > o and by definition if n = o, β is n-sound. It remains to show that (J_α, A) is 1-sound. Let $\rho = \rho_\beta^{n+1} (= \rho^1_{\alpha,A})$. If ρ = α, there's nothing to prove so suppose ρ < α. Let $p = p^1_{\alpha,A}$, and let $X = h_{\alpha,A}"(\omega\times\omega\rho\times\{p\})$. Thus $\omega\rho \cup \{p\} \subseteq X$, $(X, A\cap X) \prec_1 (J_\alpha, A)$. Let $\pi: (J_{\bar{\alpha}}, \bar{A}) \xrightarrow{\sim} (X, A\cap X)$, and let $\pi(\bar{p}) = p$. Thus $(J_{\bar{\alpha}}, \bar{A})$ is amenable, $\rho \le \bar{\alpha}$ and $J_{\bar{\alpha}} = h_{\bar{\alpha},\bar{A}}"(\omega\times\omega\rho\times\{\bar{p}\})$. Also $\bar{p} \le_J p$ and $\pi|\omega\rho = id|\omega\rho$. Let $a \in \mathcal{P}(\omega\rho)\setminus J_\alpha$ be such that a is $\Sigma_1(J_\alpha, A)$ in p. By a) for n, $a \notin J_\beta$. Also, a is $\Sigma_1(J_{\bar{\alpha}}, \bar{A})$ in $\bar{p}$ by the same Σ_1-formula, and $a \notin J_{\bar{\alpha}}$. If n = o, $\bar{A} = A = \emptyset$ and so $a \in J_{\bar{\alpha}+1}$. But then $\bar{\alpha} = \alpha$ since $a \notin J_\alpha$, and so $J_\alpha = h_\alpha"(\omega\times\omega\rho\times\{\bar{p}\})$. By (2.3) this means that $p \le_J \bar{p}$ so that, in fact, $p = \bar{p}$; i.e. J_α is 1-sound.

If n > o, let $n = \bar{n}+1$. By d) for $\bar{n}$, let $\tilde{f} \supseteq \pi$ and $\bar{\beta}$ be such that $\tilde{f}: J_{\bar{\beta}} \to_{n+1} J_\beta$, $\bar{\alpha} = \rho^n_{\bar{\beta}}$, $\bar{A} = A^n_{\bar{\beta}}$ and with the other properties of d). By a) for n and $\bar{\beta}$ it follows that a is $\underset{\sim}{\Sigma}_{n+1}(J_{\bar{\beta}})$ but $a \notin J_{\bar{\beta}}$. Accordingly, $a \in J_{\bar{\beta}+1}$, and since $a \notin J_\beta$, $\bar{\beta} = \beta$. But then $\bar{\alpha} = \rho^n_{\bar{\beta}} = \rho^n_\beta = \alpha$, $\bar{A} = A^n_{\bar{\beta}} = A^n_\beta = A$. Thus $J_\alpha = h_{\alpha,A}"(\omega\times\omega\rho\times\{\bar{p}\})$ and so as before $p \le_J \bar{p}$ which means $p = \bar{p}$. Again this means that (J_α, A) is 1-sound.

STEP 2: Let $p = p^{n+1}_\beta = p^1_{\alpha,A}$. Let $f: (J_{\bar{\rho}}, \bar{B}) \to_m (J_\rho, B)$ where $(J_{\bar{\rho}}, \bar{B})$ is amenable. Let Y = range f. Thus Y is closed for pairs and so setting $X = h_{\alpha,A}"(\omega\times Y\times\{p\})$, $(X, A\cap X) \prec_1 (J_\alpha, A)$. Let $f^*: (J_{\bar{\alpha}}, \bar{A}) \xrightarrow{\sim} (X, A\cap X)$. The key technical point is:

(*): X is an $\in$-end-extension of Y, i.e. if $y \in Y$, $x \in X\cap y$ then $x \in Y$.

Proof of (*): Let $y \in Y$, $x \in X \cap y$. Let $x = h_{\alpha,A}(i,z,p)$, where $z \in Y$. Then, letting j be such that the jth Σ_1 formula is the definition of $h_{\alpha,A}$, x is the unique element of J_α such that $(j,x,i,z) \in B$. Let $f(\bar{y}) = y$, $f(\bar{z}) = z$. Since $f: (J_{\bar{\rho}},\bar{B}) \to_0 (J_\rho,B)$ and since $(J_\rho,B) \models (\exists x \in y)((j,x,i,z) \in B)$, which is a Σ_0 statement, $(J_{\bar{\rho}},\bar{B}) \models (\exists x \in \bar{y})((j,x,i,\bar{z}) \in \bar{B})$. Let $\bar{x} \in J_{\bar{\rho}}$ be such that $\bar{x} \in \bar{y}$ and $(j,\bar{x},i,\bar{z}) \in \bar{B}$. Let $x' = f(\bar{x})$. Then $x' \in y$, $(j,x',i,z) \in B$, so $x' = x$, i.e. $x \in Y$.

Now it's an easy consequence of (*) that $f^* \supseteq f$, and of course $f^*: (J_{\bar{\alpha}},\bar{A}) \to_1 (J_\alpha,A)$. Note that $J_{\bar{\alpha}} = h_{\bar{\alpha},\bar{A}}"(\omega\times\omega\bar{\rho}\times\{\bar{p}\})$, where $f^*(\bar{p}) = p$. Thus $\rho^1_{\bar{\alpha},\bar{A}} \leq \bar{\rho}$, $p^1_{\bar{\alpha},\bar{A}} \leq_J \bar{p}$.

Further $\bar{B} = \{(i,\vec{x}) \in J_{\bar{\rho}}: (J_{\bar{\alpha}},\bar{A}) \models \phi_i(\vec{x},\bar{p})\}$. This is easily seen as follows. Let $(i,\vec{\bar{x}}) \in \bar{B}$, $\vec{x} = f(\vec{\bar{x}})$. Then $(i,\vec{x}) \in B$, that is $(J_\alpha,A) \models \phi_i(\vec{x},p)$. Since $f^*: (J_{\bar{\alpha}},\bar{A}) \to_1 (J_\alpha,A)$ and $f^* \supseteq f$, $(J_{\bar{\alpha}},\bar{A}) \models \phi_i(\vec{\bar{x}},\bar{p})$. On the other hand, if $(J_{\bar{\alpha}},\bar{A}) \models \phi_i(\vec{\bar{x}},\bar{p})$ with $\vec{\bar{x}} \in J_{\bar{\rho}}$, then for the same reasons as above, $(J_\alpha,A) \models \phi_i(\vec{x},p)$, where $\vec{x} = f(\vec{\bar{x}})$. But then $(i,\vec{x}) \in B$, so $(i,\vec{\bar{x}}) \in \bar{B}$.

Suppose now that $\bar{\eta} < \bar{\rho}, \vec{\bar{x}} \in J_{\bar{\alpha}}$, ϕ is Σ_1 and $\bar{a} \subseteq \omega.\bar{\eta}$ is such that $\bar{a} = \{y: (J_{\bar{\alpha}},\bar{A}) \models \phi(y,\vec{\bar{x}})\}$. Let $\vec{x} = f^*(\vec{\bar{x}})$, $\eta = f(\bar{\eta})$. Let $\vec{\bar{x}}$ be $h_{\bar{\alpha},\bar{A}}(i,\bar{\xi},\bar{p})$ where $i < \omega$, $\bar{\xi} < \omega.\bar{\rho}$. Let j be such that the jth Σ_1 formula in $i,\bar{\xi},\bar{p}$ is: $(\exists\vec{x})(\vec{x} = h_{\bar{\alpha},\bar{A}}(i,\bar{\xi},\bar{p}) \wedge \phi(y,\vec{x}))$.

Then $\bar{a} = \{y: (j,i,\bar{\xi},y) \in \bar{B}\}$; i.e. $\bar{a}$ is such that $(\{(j,i,\bar{\xi})\}\times\omega.\bar{\eta}) \cap \bar{B} = \{(j,i,\xi)\}\times\bar{a}$. Since $(J_{\bar{\rho}},\bar{B})$ is amenable, $\bar{a} \in J_{\bar{\rho}}$. This proves:

$$(**):\quad \bar{\rho} = \rho^1_{\bar{\alpha},\bar{A}}\ .$$

Once it's proved that $\bar{p} = p^1_{\bar{\alpha},\bar{A}}$ it will follow that $\bar{B} = A^1_{\bar{\alpha},\bar{A}}$. I'll prove this at the same time as I construct $\bar{\beta}$, $\tilde{f}$. The last step will be to verify that $\tilde{f}$ has the properties of d) and to verify its uniqueness. The n+1-soundness of $\bar{\beta}$ will then be clear.

If $n = 0$, then $\alpha = \beta$, $A = \bar{A} = \emptyset$. Set $\bar{\beta} = \bar{\alpha}$, $\tilde{f} = f^*$. By c) for $n = 0$, $J_{\bar{\alpha}}$ is 1-sound. Accordingly, there is $\bar{\xi} < \omega\bar{\rho}$, $i < \omega$ such that if $\bar{q} = p^1_{\bar{\alpha}}$, then $\bar{p} = h_{\bar{\alpha}}(i,\bar{\xi},\bar{q})$. Let $q = f^*(\bar{q})$, $\xi = f(\bar{\xi})$; then $p = h_\alpha(i,\xi,q)$, which means that $p \leq_J q$. But then $\bar{p} \leq_J \bar{q}$, and so $\bar{p} = \bar{q} = p^1_{\bar{\alpha}}$. Then $\tilde{f}$ is as required.

Now suppose $n > 0$; let $n = \bar{n}+1$. By d) for $\bar{n}$, f^*, let $\tilde{f},\bar{B}$ be unique such that $\tilde{f} \supseteq f^*$, $\bar{\alpha} = \rho^n_{\bar{\beta}}$, $\bar{A} = A^n_{\bar{\beta}}$ and for $k \leq n$, $\tilde{f}(\bar{p}_k) = p_k$, $\tilde{f}|J_{\bar{\rho}_k} : (J_{\bar{\rho}_k},\bar{A}_k) \to_{k+1} (J_{\rho_k},A_k)$ where for such k, $k' = n-k$ and $\bar{p}_k = p^{k'}_{\bar{\beta}}$, $p_k = p^{k'}_{\beta}$, $\bar{\rho}_k = \rho^{k'}_{\bar{\beta}}$, $\rho_k = \rho^{k'}_{\beta}$, $\bar{A}_k = A^{k'}_{\bar{\beta}}$, $A_k = A^{k'}_{\beta}$. So in particular by c) for n, $J_{\bar{\beta}}$ is n+1-sound, i.e. $(J_{\bar{\alpha}},\bar{A})$ is 1-sound. The proof that $\bar{p} = p^1_{\bar{\alpha},\bar{A}}$ is then essentially as above for $n = 0$. Again, this means that $\bar{p} = p^{n+1}_{\bar{\beta}}$, $\bar{B} = A^1_{\bar{\alpha},\bar{A}} = A^{n+1}_{\bar{\beta}}$. Thus $f^*(p^{n+1}_{\bar{\beta}}) = p^{n+1}_{\beta}$.

By (2.6), $\bar{\alpha},\bar{A}$ are unique such that $\bar{\rho} = \rho^1_{\bar{\alpha},\bar{A}}$, $\bar{B} = A^1_{\bar{\alpha},\bar{A}}$; the uniqueness of f^* such that $f^* \supseteq f$, $f^*(p^1_{\bar{\alpha},\bar{A}}) = p^1_{\alpha,A}$, $f^*:(J_{\bar{\alpha}},\bar{A}) \to_1 (J_\alpha,A)$ then follows easily by the 1-soundness of $(J_{\bar{\alpha}},\bar{A})$, (J_α,A). By (2.7) $f^*: (J_{\bar{\alpha}},\bar{A}) \to_{m+1} (J_\alpha,A)$. If $n = 0$ this completes the proof of d). If not, let $n = \bar{n}+1$. It follows from d) for $\bar{n}$, f^* that, with the above notation, for $k \leq n$ $\tilde{f}(\bar{p}_k) = p_k$, $\tilde{f}|J_{\bar{\rho}_k} : (J_{\bar{\rho}_k},\bar{A}_k) \to_{m+k} (J_{\rho_k},A_k)$. This completes the proof of d).

STEP 3: Suppose $1 \leq m < \omega$, $\rho = \rho^{n+1}_\beta$, $B = A^{n+1}_\beta$, $\alpha = \rho^n_\beta$, $A = A^n_\beta$, $x \subseteq J_\rho$. By (2.5) and the 1-soundness of (J_α,A), x is $\undertilde{\Sigma}_m(J_\rho,B)$ iff x is $\undertilde{\Sigma}_{m+1}(J_\alpha,A)$. If $n = 0$, this is as required. If not, a) for n guarantees that x is $\undertilde{\Sigma}_{m+1}(J_\alpha,A)$ iff x is

$\underset{\sim}{\Sigma}_{m+1+n}(J_\beta)$, i.e. iff x is $\underset{\sim}{\Sigma}_{m+n+1}(J_\beta)$. If $\eta < \omega\rho$, $x \subseteq S_\eta$ and $x \in J_\beta$ then by a) for n, $x \in J_\alpha$, and so by (2.3), $x \in J_\rho$.

STEP 4: The first two assertions of b) are immediate from the definitions of p_β^{n+1} and ρ, and from a). By c) for n, (J_α,A) is sound; but $h_{\alpha,A}$ is $\Sigma_1(J_\alpha,A)$ and so $h_{\alpha,A}$ is $\Sigma_{1+n}(J_\beta)$, i.e. $\Sigma_{n+1}(J_\beta)$. If n = o and $\alpha = \beta$, this is as required. If not, b) for n guarantees there is f which is $\underset{\sim}{\Sigma}_n(J_\beta)$ and maps a subset of $\omega\alpha$ onto J_β. Let $h(i,x) = h_{\alpha,A}(i,x,p)$. Then $f \circ h$ is $\underset{\sim}{\Sigma}_{n+1}(J_\beta)$ and maps a subset of $\omega \times J_\rho$ onto J_β. Composing with a $\underset{\sim}{\Sigma}_1(J_\rho)$ map σ of $\omega.\rho$ onto $\omega{\times}J_\rho$, $f \circ h \circ \sigma$ is a map of a subset of $\omega.\rho$ onto J_β. Since σ is $\underset{\sim}{\Sigma}_1(J_\rho)$, σ is also $\underset{\sim}{\Sigma}_1(J_\beta)$. That ρ is minimal for the existence of such a map follows from an easy generalization of the diagonal argument of (2.3)a); this generalization has for hypotheses the existence of g, a map from a subset of $\omega.\eta$ onto J_β which is $\underset{\sim}{\Sigma}_{n+1}(J_\beta)$ where $\eta \leq \rho$, and the fact that ρ is minimal with $x \in P(\omega.\rho)\backslash J_\beta$ which is $\underset{\sim}{\Sigma}_{n+1}(J_\beta)$. This completes the proof of STEP 4 and of (2.9).

(2.10) It seems in order to tie up a few loose ends and to make a few remarks about the proof of (2.9).

A. It seems worthwhile to unravel the inductive proof of (2.9) and to give explicitly the definition of the g as guaranteed by b) and the $\tilde{f}$ as guaranteed by d).

Concerning g, let ρ_k , A_k , p_k be as in d) for β and for $k \leq n+1$, and let $h_k = h_{\rho_k,A_k}(-,-,p_k)$. Let σ be a $\underset{\sim}{\Sigma}_1(J_\rho)$ map of a subset of $\omega.\rho$ onto $\omega{\times}J_\rho$. Then g can be taken to be $h_n \circ \ldots \circ h_0 \circ \sigma$. Concerning $\tilde{f}$, let $\bar{\rho}_k$, $\bar{A}_k$, $\bar{p}_k$, $\bar{h}_k$ be as above but for $\bar{\beta}$ (once $J_{\bar{\beta}}$ is constructed!). Then $\tilde{f}$ just makes the diagram commute, i.e. $\tilde{f} \circ \bar{h}_n \circ \ldots \circ \bar{h}_0 = h_n \circ \ldots \circ h_0 \circ f$. More accurately, let X_0 = range f and for $k \leq n$, let $X_{k+1} = h_{k+1}"(\omega{\times}X_k)$. Then, letting $\pi\colon X_{k+1} \xrightarrow{\sim} J_{\gamma_{k+1}}$, $\gamma_{k+1} = \bar{\rho}_{k+1}$,

$\tilde{f}|J_{\bar{\rho}_{k+1}} = \pi^{-1}$.

B. The following is a good list of rather easy exercises.

1. Let $n \geq 1$, $\eta < \rho^n_\beta$. Prove that if f is a map from a subset of $\omega\eta$ into $\omega\rho^n_\beta$ and f is $\underset{\sim}{\Sigma}_n(J_\beta)$ then f is not onto; that is, $\omega\rho^n_\beta$ is a $\underset{\sim}{\Sigma}_n(J_\beta)$-cardinal.

2. Let $n \geq 1$. Then $\omega\rho^n_\beta$ is Σ_n-admissible (though it may not be even A^n_β-admissible).

3. Note that it may be that $\rho^n_\beta < \beta$ and $J_\beta \models \omega.\rho^n_\beta$ is singular. Also there may be $\eta < \rho^n_\beta$ and f, a map from a subset of $\omega\eta$ cofinally into $\omega\beta$ which is $\underset{\sim}{\Sigma}_1(J_\beta)$.

4. Prove that J_β is Σ_n-uniformizable (hint: let $n \geq 2$ and R be a $\underset{\sim}{\Sigma}_n(J_\beta)$ binary relation. Find $\bar{R}$ a $\underset{\sim}{\Sigma}_1(J_\rho,A)$ binary relation on J_ρ such that the uniform Σ_1-uniformization of $\bar{R}$ lifts to a (no longer uniform, because of the p^k_β for $1 \leq k \leq n-1$) $\underset{\sim}{\Sigma}_n$-uniformization of R, where $\rho = \rho^{n-1}_\beta$, $A = A^{n-1}_\beta$).

(2.11) In this paragraph I prove the basic "map-splitting" lemma which will play an important role in the construction of morasses in L. Strictly speaking, the only reason to consider this lemma as part of the fine structure theory is the use of the J_α-hierarchy (which can be avoided by certain artifices) and the presence of "effective" notions like Σ_0 and Σ_1-elementary embeddings. On the other hand, the underlying intuitions most certainly do come from the fine structure theory as the notation below suggests. Typically, there are β_0,β_2, n such that ν_i is a $\underset{\sim}{\Sigma}_n$ but not a $\underset{\sim}{\Sigma}_{n+1}$ cardinal in J_{β_i}, $\rho_i = \rho^n_{\beta_i}$ $(i = 0,2)$.

<u>LEMMA</u>: Let ρ_0,ρ_2,A_0,A_2 be such that either $A_0 = A_2 = \emptyset$ or ρ_0,ρ_2 are admissible, and (J_{ρ_i},A_i) are amenable $(i = 0,2)$. Let $\nu_0,\nu_1,\nu_2,f,g_0,g_1$ be such that

i) ν_i is a regular cardinal of J_{ρ_i} $(i = 0,2)$,

ii) $f:(J_{\rho_0},A_0) \to_1 (J_{\rho_2},A_2), f|J_{\nu_0} = g_1 \circ g_0, g_0:J_{\nu_0} \to_{0,\text{cofinal}} J_{\nu_1}$,

$g_1:J_{\nu_1} \to_0 J_{\nu_2}$,

THEN, there are ρ_1,A_1,f_0,f_1 such that (J_{ρ_1},A_1) is amenable, ν_1 is a regular cardinal of J_{ρ_1}, $f_0:(J_{\rho_0},A_0) \to_{0,\text{cofinal}} (J_{\rho_1},A_1)$, $f_1:(J_{\rho_1},A_1) \to_0 (J_{\rho_2},A_2)$, $g_i = f_i|J_{\nu_i}$ $(i = 0,1)$, $f = f_1 \circ f_0$.

Proof: The proof proceeds by expressing (J_{ρ_i},A_i) $(i = 0,2)$ as directed unions of Σ_0-elementary substructures in the same way. These substructures collapse down to elements of J_{ν_i} $(i = 0,2)$. In this way the directed union systems become directed, commutative systems of models and maps which constitute $\Sigma_1(J_{\rho_i},A_i)$ predicates $\Omega_i \subseteq J_{\nu_i}$ $(i = 0,2)$ in the same Σ_1-defining formula. The g_1-inverse image of Ω_2 is a directed, commutative system of models and maps $\Omega_1 \subseteq J_{\nu_1}$, by the Σ_0-elementarity of g_1. Completing the diagram, the direct limit through Ω_1 maps into (J_{ρ_2},A_2) and hence this direct limit is well-founded. Taking it to be a transitive $\in$-model yields (J_{ρ_1},A_1) and the map into (J_{ρ_2},A_2) is f_1. f_0 is just $f_1^{-1} \circ f_0$, and is also the completion of the diagram between the direct limit through Ω_0 and the direct limit through Ω_1. If f_0 is not cofinal, since f_1 need only be Σ_0-elementary, $\omega\rho_1$ can be redefined to be the point into which f_0 is cofinal.

If ρ_0,ρ_2 are successor ordinals (which means $A_0 = A_2 = \emptyset$), the above account is slightly inaccurate. Letting $\rho_i = \gamma_i+1$ $(i = 0,2)$, instead of constructing directly ρ_1,f_0,f_1, I construct γ_1, and $\bar{f}_i: J_{\gamma_i} \to_\omega J_{\gamma_{i+1}}$, and then use the standard facts that $f|J_{\gamma_0} : J_{\gamma_0} \to_\omega J_{\gamma_2}$ and that there are canonical extensions, $f_i \supseteq \bar{f}_i$, $f_i : J_{\gamma_i+1} \to_0 J_{\gamma_{i+1}+1}$, $f_i(J_{\gamma_i}) = J_{\gamma_{i+1}}$. f_i is the extension of $\bar{f}_i$ by rudimentary functions, i.e., for $\vec{x} \in J_{\gamma_i}$, $f_i(h(\vec{x},J_{\gamma_i})) = h(\bar{f}_i(\vec{x}),J_{\gamma_{i+1}})$, where h is rudimentary. The Σ_ω-elementarity of $\bar{f}_i$ guarantees that this is unambiguous and that f_i so defined extends $\bar{f}_i$ and is Σ_0-elementary.

The $\bar{f}_i$ are obtained in essentially the same way as above. The J_{γ_i} $(i = 0,2)$ are expressed in the same way as directed unions of Σ_n-elementary substructures $(n < \omega)$. The collapsed systems of models and Σ_n-elementary maps form Σ_1 predicates $\Omega_i \subseteq J_{\nu_i}$ defined by the same Σ_1 formula over J_{ρ_i} $(i = 0,2)$. J_{γ_1} is obtained as the direct limit through Ω_1-the directed, commutative system of models and Σ_n-elementary maps $(n < \omega)$ which is the g_1-inverse image of Ω_2. Once again, $\bar{f}_1$ is obtained by completing the diagram and $\bar{f}_0 = \bar{f}_1^{-1} \circ f|J_{\gamma_0}$. This said, I'll just describe the directed union systems and the directed commutative systems of models and maps Ω_i $(i = 0,2)$ in two cases.

<u>CASE 1</u>: ρ_i are limit ordinals $(i = 0,2)$.

First, recall that $h_{M,B}$ is defined and lacks only the definability over (M,B) when (M,B) is non-amenable and even then $h_{M,B}$ is rudimentary in M and B. Hence, if $\alpha < \rho_i, h_{\alpha,A_i \cap \alpha} \in J_{\rho_i}$; in what follows, h^i_α denotes $h_{\alpha,A_i \cap J_\alpha}$.

Set $t \in T_i$ iff $t = (\eta,\xi,a)$ where $\eta < \rho_i$, $a \in J_\eta$ is finite and $\xi < \nu_i$ $(i = 0,2)$. For such $t \in T_i$ let $X_t = h^i_\eta{}''(\omega\times\omega.\xi\times\{a\})$. Thus, by the above, $X_t \in J_{\rho_i}$. For $t,t' \in T_i$, let $t \precsim t'$ iff $\eta \le \eta'$, $\xi \le \xi'$, $a \subseteq a'$; thus, if $t \precsim t'$ then $X_t \subseteq X_{t'}$. Clearly, since ρ_i is a limit ordinal, $J_{\rho_i} = \bigcup_{t\in T_i} X_t$. This is the directed union system. Also $(T_i, \prec)$ is $\Sigma_1(J_{\rho_i}, A_i)$ by the same Σ_1 formula, as is the function $((X_t, A_i \cap X_t): t \in T_i)$.

Now, for $t \in T_i$, let $\sigma_t: (J_{\eta_t^*}, A_t^*) \xrightarrow{\sim} (X_t, A_i \cap X_t)$, let a_t^* be $\sigma_t^{-1}(a)$. Since $\eta_t^* \subseteq \eta$, if $A_i = \emptyset$, $(J_{\eta_t^*}, A_t^*) \in J_{\rho_i}$, while if $A_i \neq \emptyset$, ρ_i is admissible so that σ_t and hence $A_t^* \in J_{\rho_i}$. In what follows, let $\mathcal{O}\!\!\mathcal{l}_t = (J_{\eta_t^*}, A_t^*)$, and let $h^t = h_{\mathcal{O}\!\mathcal{l}_t}$. Then $|\mathcal{O}\!\!\mathcal{l}_t| = h^t{}''(\omega\times\omega.\xi\times\{a_t^*\})$, and $h^t \in J_{\rho_i}$. Since $\xi < \nu_i$ and ν_i is a regular cardinal of J_{ρ_i}, this means that $\eta_t^* < \nu_i$, and, if $A_i \neq \emptyset$, $A_t^* \in J_{\nu_i}$, since the GCH condensation argument can then be carried out in J_{ρ_i}.

If $t \prec t'$ let $\sigma_{tt'} = \sigma_{t'}^{-1} \circ \sigma_t : \mathcal{O}\!\!\mathcal{l}_t \to_0 \mathcal{O}\!\!\mathcal{l}_{t'}$. Also, $\sigma_{tt'}$ is just the inverse of the transitive collapse of $\sigma_{t'}^{-1}[X_t]$ and $\sigma_{t'}^{-1}[X_t]$ is just $h^{t'}{}''(\omega\times\omega.\xi\times\{\sigma_{t'}^{-1}(a)\})$. Since J_{ν_i} is admissible (ν_i is a regular cardinal of J_{ρ_i}), $\sigma_{tt'} \in J_{\nu_i}$.

Let $\Omega_i = ((\mathcal{O}\!\!\mathcal{l}_{t'}, \sigma_{tt'}): t, t' \in T_i, t \prec t')$. It's not hard to see that Ω_i is $\Sigma_1(J_{\rho_i}, A_i)$ in ν_i by the same Σ_1 formula. Let $\Omega_1 = g_1^{-1}[\Omega_2]$. Then Ω_1 is a directed, commutative system of models and Σ_0-elementary maps. Let $(((M,E,\bar{A}), \bar{\sigma}_t): t \in \bar{T})$ be the direct limit through Ω_1. Thus, $\bar{T} = \{t \in T_2 : \mathcal{O}\!\!\mathcal{l}_t \in \text{range } g_1\}$. Let $\bar{f}: (M,E,\bar{A}) \to_0 (J_{\rho_2}, A_2)$ be given by $\bar{f} \circ \bar{\sigma}_t = \sigma_t \circ g_1$ $(t \in \bar{T})$. Thus $(M,E,\bar{A})$ is well-founded and extensional. Also $(M,E,\bar{A}) \models$ "I am a J_α" (this is easily checked). Thus, taking (M,E) to be a transitive ϵ-model, ρ_1, A_1 are defined by taking $(J_{\rho_1}, A_1) = (M,E,\bar{A})$, in which case f_1 is just $\bar{f}$. Clearly range $f \subseteq$ range f_1.

<u>CASE 2</u>: ρ_i are successor ordinals, say $\rho_i = \gamma_i+1$ $(i = 0,2)$.

Then $f(J_{\gamma_0}) = J_{\gamma_2}$ and so $\bar{f} = f|J_{\gamma_0} : J_{\gamma_0} \to_\omega J_{\gamma_2}$. Let $t \in T_i$ iff $t = (n,\xi,a)$ where $n < \omega$, $\omega.\xi < \nu_i$, $a \in J_{\gamma_i}$ is finite. For such t, let X_t be the set of members of J_{γ_i} which are the $<_J$-least members of J_{γ_i} satisfying a $\Sigma_n(J_{\gamma_i})$ condition in parameters from $\omega.\xi \cup a$. Thus $X_t \prec_n J_{\gamma_i}$ and $J_{\gamma_i} = \bigcup_{t \in T_i} X_t$. Set $t \prec t'$ iff $n \leq n'$, $\xi \leq \xi'$, $a \subseteq a'$; thus, if $t \prec t'$ then $X_t \prec_n X_{t'}$. Now proceed as in CASE 1 (changing a few details) to obtain γ_1, $\bar{f}_0, \bar{f}_1$, and thence, as outlined above, to obtain $\rho_1 = \gamma_1+1$ and f_0, f_1.

§3 MORASSES AND CONSTRUCTING THEM IN L.

In this section I'll work gradually up to defining a morass and constructing one in L. First, I'll construct in L a simpler object called a coarse morass which is then defined after the fact by abstraction from the properties of the object which was constructed. A morass is then defined to be a coarse morass which satisfies two further more subtle properties which Jensen aptly called continuity properties. Finally, I (considerably) modify the coarse morass construction to give a morass in L.

As mentioned in the introduction, I assume V = L, but of course this is needed only to facilitate the presentation of the constructions and plays no real role (other than motivation) in the definition of morass. To fix ideas, I'll construct an ω_1 (coarse) morass, but the construction generalizes readily to arbitrarily uncountable regular κ, giving κ(coarse) morasses. The definitions are given in this full generality. A few simple and general observations will help get under way.

(3.1) PROPOSITION: Suppose $\alpha < \nu$ and $p \in L_\nu$ is such that every element of L_ν is definable in L_ν in parameters from $\alpha \cup \{p\}$. Then

a) the set of elementary substructures of L_ν containing p and having transitive intersection with α is well-ordered by inclusion and the well-ordering coincides with the ordering of the intersections with α,

b) the set of elementary substructures of L_ν containing p and having transitive intersection with α is closed under unions of chains.

Proof: b) is evident. For a) note that if $p \in X \prec_\omega L_\nu$ and $X \cap \alpha = \beta$ then $X = \{(t(\xi,p))^{L_\nu}: \xi < \beta$, t is an L_ν-term$\}$.

(3.2) REMARK: Proposition (3.1) holds if "definable" is replaced by "Σ_1-definable" and "elementary substructures" is replaced by "s-elementary substructures" where s is a reasonably closed set of formulas containing all the Σ_1-formulas. The proof of a) in this setting proceeds simply by restricting to Σ_1-terms t. Also, L_ν can be replaced by J_ν or by structures of the form (L_ν,A), (J_ν,A). τ will be called non-projectible (because then $\tau = \rho^n_\tau$ for all n) if $L_\tau \models ZF^-$; if so, of course $L_\tau = J_\tau$.

(3.3) PROPOSITION: Suppose τ is non-projectible or a limit of non-projectible ordinals, that $\omega < \alpha < \rho < \tau$, $L_\tau \models \alpha$ is regular, $p \in J_\rho$, $A \in P(J_\rho) \cap L_\tau$. Then:

a) if $X \prec_0 L_\tau$, $p, \rho, A \in X$ then $(X \cap J_\rho, X \cap A) \prec_\omega (J_\rho, A)$,

b) there is an increasing continuous tower of length α consisting of elementary substructures of (J_ρ,A) which contain p, have transitive intersection with α and lie in L_τ.

Proof: Suppose first that τ is non-projectible. a) is clear and standard by relativation; b) is easy since using the closure properties of L_τ and the fact that $L_\tau \models \alpha$ is regular, such a tower can be constructed working inside L_τ. Now if τ is a limit of non-projectible ordinals isolate everything inside some $L_{\tau'}$, where $\tau' < \tau$ is non-projectible.

I'll now begin constructing the coarse morass.

(3.4) DEFINITION: ν is ω_2-like iff ν is non-projectible and there is exactly one α which is an uncountable regular cardinal of L_ν; this α is denoted by α_ν so that $\alpha_\nu = \omega_1^{L_\nu}$. ν is pretty iff ν is ω_2-like and there is $p \subseteq \alpha_\nu$, $p \in L_\nu$ such that all elements of L_ν are L_ν-definable in parameters from $\alpha_\nu \cup \{p\}$; the $<_L$-least such p is denoted by p_ν.

(3.5) DEFINITION: Let $S^1 = \{\nu: \nu \text{ is pretty}\}$, $S^0 = \{\alpha_\nu: \nu \text{ is pretty}\}$,

$$\mathcal{S} = \{(\alpha_\nu,\nu): \nu \text{ is pretty}\}. \quad \text{For } \alpha \in S^0, \text{ let } S_\alpha = \{\nu: \alpha = \alpha_\nu\}.$$

(3.6) PROPOSITION: (MO)

a) $\mathcal{S}$ is a set of pairs of **primitive-recursively (p.r.)** closed ordinals such that if $(\alpha,\nu) \in \mathcal{S}$, $\alpha < \nu < \omega_2$,

b) if $(\alpha,\nu),(\alpha',\nu') \in \mathcal{S}$ and $\alpha < \alpha'$ then $\nu < \alpha'$,

c) $\omega_1 = \max S^0 = \sup S^0 \cap \omega_1$, $\omega_2 = \sup S^1 = \sup S_{\omega_1}$.

Proof: Clear.

(3.7) DEFINITION: If $\nu_1,\nu_2 \in S^1$, $f: L_{\nu_1} \to L_{\nu_2}$ is pretty iff f is an elementary embedding and $f(p_{\nu_1}) = p_{\nu_2}$.

(3.8) PROPOSITION:

a) If $f: L_{\nu_1} \to L_{\nu_2}$ is pretty then $f|\alpha_{\nu_1} = id|\alpha_{\nu_1}$, $f(\alpha_{\nu_1}) = \alpha_{\nu_2}$.

b) There is at most one pretty $f: L_{\nu_1} \to L_{\nu_2}$.

c) If $\nu_1 \neq \nu_2$ and $f: L_{\nu_1} \to L_{\nu_2}$ is pretty then $\alpha_{\nu_1} < \alpha_{\nu_2}$.

<u>Proof</u>: For a) it's clear that $f(\alpha_{\nu_1}) = \alpha_{\nu_2}$ since α_{ν_1} is the ω_1 of L_{ν_1}. The proof that $f|\alpha_{\nu_1} = id|\alpha_{\nu_1}$ is by induction on $\beta \le \alpha_{\nu_1}$. If $\beta \le \omega$ then clearly $f(\beta) = \beta$, and if $f(\beta) = \beta$ then $f(\beta+1) = \beta+1$. So let $\beta > \omega$ be a limit ordinal and suppose that $f|\beta = id|\beta$. Since $\beta < \alpha_{\nu_1}$, there is, in L_{ν_1}, $g:\omega \to_{onto} \beta$. By elementarity $f(g):f(\omega) \to_{onto} f(\beta)$, and for $n < \omega$ $f(g)(f(n)) = f(g)(n) = f(g(n)) =$ (since $f|\beta = id|\beta$ and $g(n) < \beta$) $g(n)$. Hence $f(g) = g:\omega \to_{onto} f(\beta)$ and so $f(\beta) = \beta$.

For b), if $f_1,f_2: L_{\nu_1} \to L_{\nu_2}$ are pretty, then for L_{ν_1}-terms t, for $\xi < \alpha_{\nu_1}$, $t(\xi,p_{\nu_2})$ is defined in L_{ν_2} and $f_1((t(\xi,p_{\nu_1}))^{L_{\nu_1}}) = t(\xi,p_{\nu_2})^{L_{\nu_2}} = f_2((t(\xi,p_{\nu_1}))^{L_{\nu_2}})$.

For c), clearly $\alpha_{\nu_1} \le \alpha_{\nu_2}$. If $\alpha_{\nu_1} = \alpha_{\nu_2}$ then L_{ν_1}, L_{ν_2} are isomorphic to the term models on $\alpha = \alpha_{\nu_1} = \alpha_{\nu_2}$ which are given by the complete theories of $(L_{\nu_1},p_{\nu_1},\xi)_{\xi<\alpha}$, $(L_{\nu_2},p_{\nu_2},\xi)_{\xi<\alpha}$. By the elementarity of f, these theories coincide, hence so do the term models. Since L_{ν_1}, L_{ν_2} are both transitive and isomorphic to this term model, they're equal.

(3.9) <u>DEFINITION</u>: For $\nu_1,\nu_2 \in S^1$, set $\nu_1 \; T \; \nu_2$ iff there is pretty $f: L_{\nu_1} \to L_{\nu_2}$. If $\nu_1 \; T \; \nu_2$ let $f_{\nu_1\nu_2}$ be the unique pretty $f: L_{\nu_1} \to L_{\nu_2}$; conventionally set $f_{\nu_1\nu_2}(\nu_1) = \nu_2$.

(3.10) <u>PROPOSITION</u>: (M1)

a) T is a tree on S^1 such that if $\nu_1 \; T \; \nu_2$ then $\alpha_{\nu_1} < \alpha_{\nu_2}$.

b) $(f_{\nu_1\nu_2} : \nu_1 \; T \; \nu_2)$ is a commutative system of order-preserving maps, $f_{\nu_1\nu_2}|\nu_1+1:\nu_1+1 \to \nu_2+1$, $f_{\nu_1\nu_2}(\nu_1) = \nu_2$.

c) $f_{\nu_1\nu_2}|\alpha_{\nu_1} = id|\alpha_{\nu_1}, f_{\nu_1\nu_2}(\alpha_{\nu_1}) = \alpha_{\nu_2}$.

d) setting $\alpha_i = \alpha_{\nu_i}, S_i = S_{\alpha_i} \cap \nu_i+1$ $(i = 1,2)$, then $f_{\nu_1\nu_2}: S_1 \to S_2$ and ν_1 is minimal in S_1 iff ν_2 is minimal in S_2 and if ν_1 is the immediate successor of τ_1 in S_1 and $\tau_2 = f_{\nu_1\nu_2}(\tau_1)$ then ν_2 is the immediate successor of τ_2 in S_2.

<u>Proof</u>: Everything but d) is now totally clear and d) is an easy consequence of the elementarity of $f_{\nu_1\nu_2}$.

(3.11) <u>PROPOSITION</u>: a) (M2): If $\nu_1 \, T \, \nu_2$, $\tau_1 \in S_1$ (as in (3.10)d)) and $\tau_2 = f_{\nu_1\nu_2}(\tau_1)$ then $\tau_1 T \, \tau_2$ and $f_{\tau_1\tau_2}|\tau_1+1 = f_{\nu_1\nu_2}|\tau_1+1$ (in fact, $f_{\tau_1\tau_2} = f_{\nu_1\nu_2}|L_{\tau_1}$).

b) (M3): $\{\alpha_\nu: \nu T \tau\}$ is closed in α_τ.

c) (M4): If τ is not maximal in S_{α_τ} then $\{\alpha_\nu: \nu T\tau\}$ is unbounded in α_τ.

d) (M5): If $\{\alpha_\nu: \nu T\tau\}$ is unbounded in α_τ then $\tau = \bigcup_{\nu T\tau} f_{\nu\tau}{}''\nu$.

<u>Proof</u>: Clear from (3.3).

I'll now define a coarse morass by abstraction from the preceding construction.

(3.12) <u>DEFINITION</u>: Let $\kappa > \omega$ be regular. A κ-coarse morass is a structure $(S, S^0, S^1, T, f_{\nu_1\nu_2})_{\nu_1 T \nu_2}$ satisfying (MO)-(M5) except that in (MO) "κ" replaces "ω_1" and "κ^+" replaces "ω_2".

In fact, the very notion of coarse morass and the properties (MO)-(M5) were formulated in the attempt to capture in abstract form the salient properties of the structure just constructed.

This represents an important aspect of morasses: in addition to recapitulating certain important combinatorial properties, in addition to representing a tool for carrying out intricate inductive constructions, the formulation of the notion of morasses may be thought of as a step beyond the combinatorial principles $\diamondsuit$, $\diamondsuit^+$, $\square$ towards isolating important structural features of the constructible universe.

Before going further I should elaborate a bit on the morass properties (M0)-(M5), and draw a few pictures to help the reader visualize morasses.

FIGURE 1:

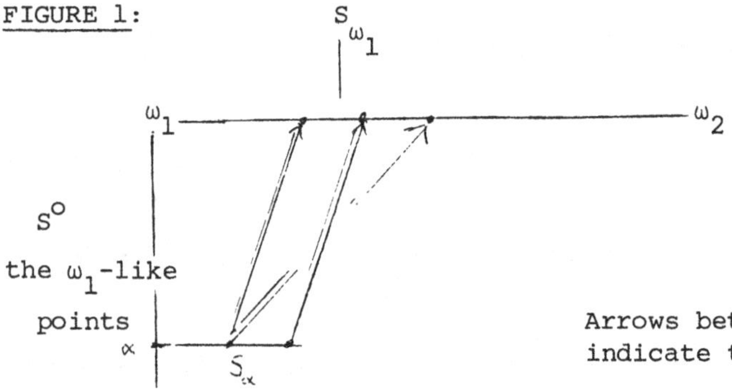

Arrows between points indicate the tree relation.

The tree T and maps $f_{\nu_1\nu_2}$ express how the points in S_{ω_1} are "approximated" by a chain of height ω_1 of countable ordinals. Provided that the process of approximation has certain coherence properties, there's hope of determining a structure of size ω_2 - in the simplest case, the interval $[\omega_1,\omega_2)$ - as a direct limit through a tree of height ω_1 consisting of countable structures. Thus, there is the obvious connection with the gap-two-two-cardinal problem of model theory (see [6], or [2]); in fact one of the major motivations for the formulation of the notion of morass was the attempt to solve this problem.

There is little to say about the properties (M0),(M1) which tell what sort of structure is being considered. The property

(M2) is the first attempt to enforce certain coherence properties on the process of approximation, in that if a point is mapped to another by a tree map, then the former stands in the tree relation to the latter with the restriction of the tree map in question as witness.

FIGURE 2:

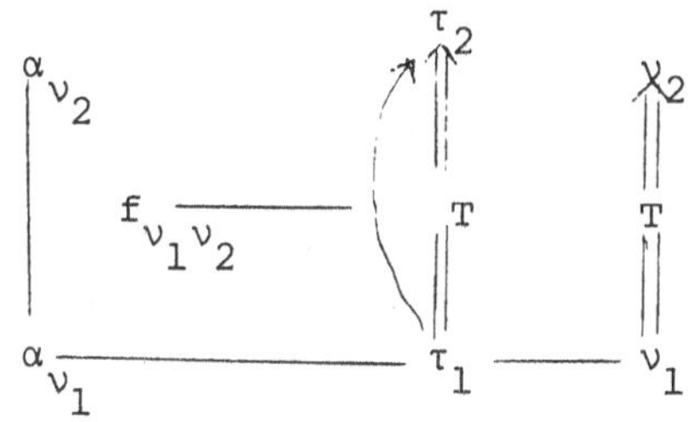

The property (M3) says, in effect, that there are no gaps in the process of approximating a point $\tau \in S^1$. The property (M4) can be thought of as being motivated from a practical point of view: in most inductive constructions, the limit stages are rather easy to handle. Thus, it would be desirable to have as many limit steps as possible in the process of approximation. A more systematic explanation runs like this: associated with each ω_1-like point α are a certain set (S_α) of ω_2-like points. All but the largest ω_2-like point associated with α (if it exists) can be expressed as a direct limit (of length α in the above construction) of structures of size smaller than α. By (M3) the conclusion of (M4) is equivalent to saying that τ is a limit in T. It is immediate from (M4) that the S_α's are <u>not</u> the levels of T, because if α is a successor point in the increasing enumeration of S^o, then S_α has exactly one element. The property (M5) merely guarantees that the process of approximation has certain natural properties, namely, a limit point in T is uniquely determined as the direct limit of the tree maps between its predecessors.

(3.13) DEFINITION: A κ-morass is a κ-coarse morass $(S, S^o, S^1, T, f_{\nu_1\nu_2})_{\nu_1 T \nu_2}$ satisfying the properties (MO)', (M1)-(M5), and (M6),(M7); the statements of (M6),(M7) follow, while (MO)' is obtained from (MO) by adding:

d) for all $\alpha \in S^o$, S_α is closed as a subset of $\sup S_\alpha$.

The properties (M6),(M7) are:

(M6): Suppose $\nu_0, \nu_2 \in S^1, \alpha_i = \alpha_{\nu_i}$ $(i = 0,2)$. If ν_0 is a limit in S_{α_0}, if $\nu_0 T \nu_2$ and $\sup f_{\nu_0\nu_2}{}''\nu_0 = \nu_1 < \nu_2$, then $\nu_0 T \nu_1$ and $f_{\nu_0\nu_1}|\nu_0 = f_{\nu_0\nu_2}|\nu_0$ (note that (MO)' guarantees $\nu_1 \in S_{\alpha_2}$ and in fact that $\nu_1 \in S^*_{\alpha_2}$).

(M7): Suppose $\bar\tau$ is a limit in $S_{\bar\alpha}$, $\bar\tau$ is the immediate T-predecessor of τ and $\tau = \sup f_{\bar\tau\tau}{}''\bar\tau$. Then, whenever $\bar\alpha < \alpha < \alpha_\tau$ and $\alpha \in S^o$, there is $\bar\nu \in S_{\bar\alpha} \cap \bar\tau$ such that setting $\nu = f_{\bar\tau\tau}(\bar\nu)$, there is no η with $\alpha_\eta = \alpha$ and $\bar\nu T \eta T \nu$.

The properties (M6) and (M7) are the most subtle, difficult to prove, and difficult to motivate. Like (M3) they are continuity properties: (M3) expresses a sort of "vertical" continuity, while (M6) expresses a sort of "horizontal" continuity, and, as will be explained, (M7) expresses a sort of "diagonal" continuity.

FIGURE 3:

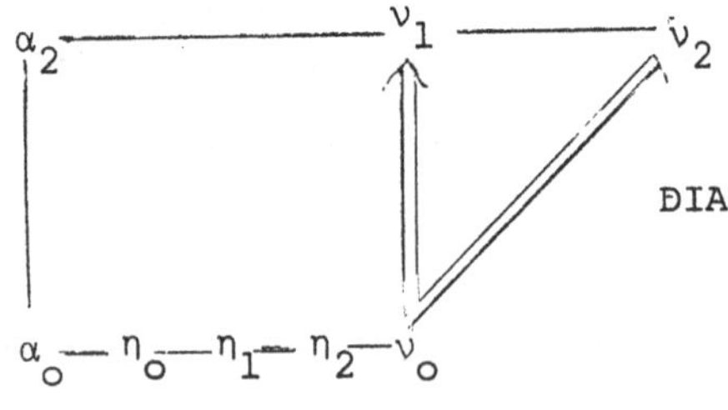

DIAGRAM FOR (M6)

Before attempting to depict the situation described in (M7) it will be useful and instructive to present an alternative formulation equivalent to (M7):

Suppose $\nu_0,\nu_2 \in S^1, \alpha_i = \alpha_{\nu_i}$ $(i = 0,2)$, ν_0 is a limit in S_{α_0}, $\nu_0 T \nu_2, \nu_2 = \sup f_{\nu_0\nu_2}{}''\nu_0$, $\alpha_0 < \alpha_1 < \alpha_2$. Suppose that for all $\eta_0 \in S_{\alpha_0} \cap \nu_0$, setting $\eta_2 = f_{\nu_0\nu_2}(\eta_0)$, there is η_1 such that $\eta_0 T \eta_1 T \eta_2$ and let η_0^* be the least such η_1. Suppose that $\alpha_1 = \sup\{\alpha_{\eta_0^*}: \eta_0 \in S_{\alpha_0} \cap \nu_0\}$ (note that by (M2),(M3) this means that $\alpha_1 \in S^0$ and in fact that if $\eta_0 \in S_{\alpha_0} \cap \nu_0$ then there is $\eta_1 \in S_{\alpha_1}$ with $\eta_0 T \eta_1 T \eta_2$). Then there is $\nu_1 \in S_{\alpha_1}$ such that $\nu_0 T \nu_1 T \nu_2$ and ν_0 is the immediate T-predecessor of ν_1.

This explains the above terminology: ν_1 is the diagonal limit of the η_0^*.

FIGURE 4:

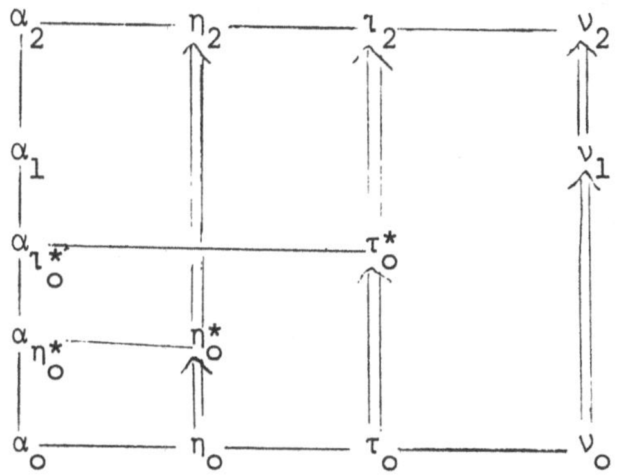

Thus (M7) again expresses the "saturatedness" of the morass in that any stage in the process of approximation which arises naturally (e.g. ν_1) is already part of the morass. A pragmatic explanation of the first formulation of (M7) is that, again, limit steps in an inductive constructions are easier to handle. Though τ is an immediate T-successor of $\bar{\tau}$, it can be viewed as a diagonal limit (of the η_0^* of the second formulation).

Now I'll construct a morass in L. Once again, this construction should be thought of as the motivating example behind the formulation of the notion of morass and the morass properties as attempts to capture certain important structural features of L encountered in the construction. I'll construct an ω_1-morass; the generalization to arbitrary regular $\kappa > \omega$ is straightforward, as in the generalization to L[A] when $A \subseteq \kappa$.

(3.14) <u>DEFINITION:</u> $\nu \in S^1$ iff ν is not a cardinal, ν is a limit of non-projectible ordinals, and there is a unique uncountable cardinal, α_ν in L_ν (for κ-morasses in L[A] when $A \subseteq \kappa$, we require $\nu < \kappa^+$, ν is not a cardinal, there is $\omega < \alpha_\nu < \nu$ such that $L_\nu[A \cap \alpha_\nu] \models$ "α_ν is the largest cardinal and α_ν is regular", and ν is a limit of ordinals τ such that $L_\tau[A \cap \alpha_\nu] \models ZF^-$). $S, S^0, S_\alpha (\alpha \in S^0)$ are defined as in (3.5).

(3.15) <u>DEFINITION:</u> If $\nu \in S^1$, let $\beta(\nu)$ be the least $\beta \geq \nu$ such that ν is not a cardinal in $J_{\beta+1}$.

Accordingly, letting $\beta = \beta(\nu)$, there is $k < \omega$ such that there are $p \in J_\beta$, $\eta < \nu$, $a \subseteq \eta$, and f with $f: a \to_{\text{onto}} \nu$ and f is $\Sigma_k(J_\beta)$ in p. Such a k is positive.

(3.16) <u>DEFINITION:</u> Let $\beta = \beta(\nu)$. $n(\nu)$ is the least n for the existence of p, η, a, f with $\eta < \nu$, $a \subseteq \eta$, $f: a \to_{\text{onto}} \nu$ and f is $\Sigma_{n+1}(J_\beta)$ in p. Let $n = n(\nu)$. Set $\rho(\nu) = \rho^n_\beta$ and let $A(\nu) = A^n_\beta$.

(3.17) <u>PROPOSITION:</u> Let $\nu \in S^1$, $\beta = \beta(\nu)$, $n = n(\nu)$, $\rho = \rho(\nu)$, $A = A(\nu)$ Let $\gamma = \rho^{n+1}_\beta$. THEN: $\rho \geq \nu$, $\omega.\gamma \leq \alpha_\nu$.

<u>Proof:</u> Clearly, if $n = 0$, $\rho = \beta \geq \nu$. If $n > 0$, $m < n$, and $\rho^m_\beta \geq \nu$, let $\delta = \rho^m_\beta$, $B = A^m_\beta$, $q = p^m_\beta$, $\eta = \rho^{m+1}_\beta$. Since $J_\delta = h_{\delta,B}$"$(\omega\times\omega.\eta\times\{q\})$, there is a $\underset{\sim}{\Sigma}_1(J_\delta)$ map g of a subset of $\omega.\eta$ onto ν. This contradicts the definition of $n(\nu)$, unless $\eta \geq \nu$.

On the other hand, if $f: a \to_{\text{onto}} \nu$, $a \subseteq \eta < \nu$, and f is $\Sigma_{n+1}(J_\beta)$, then f is $\Sigma_1(J_\rho, A)$. Also $f \notin J_\rho$ because $J_\rho \subseteq J_\beta$ and ν is a regular cardinal of J_β. Thus, since $\nu = \omega\nu$, $\gamma \leq \nu$. But $J_\rho = h_{\rho,A}"(\omega\times\omega\gamma\times\{p_\beta^{n+1}\})$. If $\omega\gamma = \nu$ then, setting $h = h_{\rho,A}(-,-,p_\beta^{n+1})$, $h\circ f$ maps a subset of η onto J_ρ and $\eta < \omega\gamma$: this is impossible. If $\alpha_\nu < \omega\gamma < \nu$, there is $g \in L_\nu$, $g: \alpha_\nu \to_{\text{onto}} \omega\gamma$ and so $h\circ g$ maps a subset of α_ν onto J_ρ and $\alpha_\nu < \omega\gamma$: again, this is impossible.

Note that (3.17) guarantees that $J_\rho = h_{\rho,A}"(\omega\times\alpha_\nu\times\{p\})$ where ρ, A are as above and $p = p_\beta^{n+1}$.

(3.18) DEFINITION: $p(\nu)$ is the $<_J$-least $p \in J_\rho$ such that $J_\rho = h_{\rho,A}"(\omega\times\alpha_\nu\times\{p\})$ (thus, if $\omega\gamma < \alpha_\nu$, then possibly $p <_J p_\beta^{n+1}$). Set $\mathfrak{A}_\nu = (J_\rho, A, p(\nu))$.

(3.19) DEFINITION: $Qx\theta x$ is the formula: $\forall^{\text{ordinal}}y\ \exists^{\text{ordinal}}x\ (y < x \wedge \theta(x))$. ϕ is a Q-formula iff ϕ is $Qx\theta(x)$ where θ is Σ_1. Q-elementary substructures and embeddings have the obvious meaning and are denoted by $\mathfrak{A} \prec_Q \mathfrak{B}$, $\sigma: \mathfrak{A} \to_Q \mathfrak{B}$.

(3.20) DEFINITION: Suppose $\nu_1, \nu_2 \in S^1$. Set $\nu_1 T \nu_2$ iff $\nu_1 \neq \nu_2$ but

a) $n(\nu_1) = n(\nu_2)$,

b) $\rho(\nu_1) = \nu_1$ iff $\rho(\nu_2) = \nu_2$, and

c) there is $f: \mathfrak{A}_{\nu_1} \to_1 \mathfrak{A}_{\nu_2}$ such that $f|\alpha_{\nu_1} = \text{id}|\alpha_{\nu_1}$, $f(\alpha_{\nu_1}) = \alpha_{\nu_2}$, $f(p(\nu_1)) = p(\nu_2)$, $f|L_{\nu_1}: L_{\nu_1} \to_Q L_{\nu_2}$ and if $\nu_1 < \rho(\nu_1)$ then $f(\nu_1) = \nu_2$; such a map is called pretty.

(3.21) PROPOSITION: a) If $\nu_1, \nu_2 \in S^1$, there is at most one pretty $f: \mathfrak{A}_{\nu_1} \to \mathfrak{A}_{\nu_2}$,

b) T is a tree; if $\nu_1 T \nu_2$ then $\alpha_{\nu_1} < \alpha_{\nu_2}$.

Proof: Essentially as in (3.8)b),c) and (3.10)a),b).

(3.22) DEFINITION: If $\nu_1 T \nu_2$, $f_{\nu_1\nu_2}$ is the unique pretty $f:\mathfrak{A}_{\nu_1} \to \mathfrak{A}_{\nu_2}$.

(3.23) In practice, when trying to prove that $\bar{\nu}T\nu$, it frequently is relatively straightforward to find $\bar{f},\bar{\rho},\bar{A},\bar{p}$ such that $(J_{\bar{\rho}},\bar{A})$ is amenable, $\bar{\rho} \geq \bar{\nu}$, $\bar{f}:(J_{\bar{\rho}},\bar{A}) \to_1 (J_{\rho(\nu)},A(\nu))$, $\bar{f}(\bar{p}) = p(\nu)$, $\bar{f}|\alpha_{\bar{\nu}} = \mathrm{id}|\alpha_{\bar{\nu}}, f(\alpha_{\bar{\nu}}) = \alpha_\nu$, $\bar{f}|L_{\bar{\nu}}: L_{\bar{\nu}} \to_\Omega L_\nu$, $\bar{\rho} = \bar{\nu}$ iff $\rho(\nu) = \nu$, and if not then $\bar{f}(\bar{\nu}) = \nu$. Note that this means that if $\bar{\nu} < \bar{\rho}$, then $\bar{\nu}$ is a regular cardinal of $J_{\bar{\rho}}$.

What then remains is to prove that $n(\bar{\nu}) = n(\nu)$, that $p(\bar{\nu}) = \bar{p}$, that $\rho(\bar{\nu}) = \bar{\rho}$ and that $A(\bar{\nu}) = \bar{A}$. This is usually done as follows. Using (2.9)d), there are f^*,β such that, setting $n = n(\nu)$, $\beta = \beta(\nu)$, then $\bar{\rho} = \rho^n_{\bar{\beta}}$, $\bar{A} = A^n_{\bar{\beta}}$, $f^* \supseteq \bar{f}, f^*: J_{\bar{\beta}} \to_{n+1} J_\beta$ and having the other properties of (2.9)d). Note that if $\bar{\nu} < \bar{\beta}$, then $\bar{\nu}$ is a regular cardinal of $J_{\bar{\beta}}$.

Now $J_{\bar{\rho}} = h_{\bar{\rho},\bar{A}}"(\omega\times\alpha_{\bar{\nu}}\times\{\bar{p}\})$, since this carries down, via $\bar{f}$ from $(J_{\rho(\nu)},A(\nu))$. Thus there is $\bar{g}$, a $\Sigma_1(J_{\bar{\rho}},\bar{A})$ map from a subset of $\alpha_{\bar{\nu}}$ onto $\bar{\nu}$. Such a $\bar{g}$ is then $\Sigma_{n+1}(J_{\bar{\beta}})$. Thus $\bar{\beta} = \beta(\bar{\nu})$ and $n(\bar{\nu}) \leq n$. Suppose, towards a contradiction, that $n(\bar{\nu}) < n$ and set $m = n(\bar{\nu})$. Then there is σ, a $\Sigma_{m+1}(J_{\bar{\beta}})$ map from a subset of $\alpha_{\bar{\nu}}$ onto $\bar{\nu}$. If $\bar{\nu} < \bar{\rho}$, then, since $m < n$, $\bar{\nu} < \bar{\rho} \leq \rho^{m+1}_{\bar{\beta}}$, and so σ is a $\Sigma_{m+1}(J_{\bar{\beta}})$ bounded subset of $J_{\rho^{m+1}_{\bar{\beta}}}$. This means that $\sigma \in J_{\bar{\beta}}$, which is impossible since $\bar{\nu}$ is a regular cardinal of $J_{\bar{\beta}}$. Hence there is a $\Sigma_{m+1}(J_{\bar{\beta}})$ map of a subset of $\alpha_{\bar{\nu}}$ onto $\bar{\rho} = \omega.\bar{\rho}$, and, since $m+1 \leq n$, this means, by (2.9)b), that there is a $\Sigma_n(J_{\bar{\beta}})$ map of a subset of $\alpha_{\bar{\nu}}$ onto $J_{\bar{\beta}}$. This contradicts (2.9)d) which guaranteed the minimality of $\omega.\bar{\rho}$ for this property. But then, in this case too $n(\bar{\nu}) = n$. But $\bar{\rho} = \rho(\bar{\nu})$, $\bar{A} = A(\bar{\nu})$, as required, and $\bar{p}(\bar{\nu}) \leq_J \bar{p}$. Finally, suppose that $\bar{q} <_J \bar{p}$. Then $q = \bar{f}(\bar{q}) <_J p(\nu)$. But then, if $\bar{p} \in h_{\bar{\rho},\bar{A}}''(\omega\times\alpha_{\bar{\nu}}\times\{\bar{q}\})$, $p(\nu) \in h_{\rho,A}''(\omega\times\alpha_\nu\times\{q\})$ (in fact $p(\nu) \in h_{\rho,A}''(\omega\times\alpha_{\bar{\bar{\nu}}}\times\{q\})$). This contradicts the definition of $p(\nu)$, and so $p(\bar{\nu}) = \bar{p}$.

In what follows, the above schema will be used in different situations, and sometimes merely cited to complete proofs.

(3.24) <u>LEMMA</u>: $(S, S^0, S^1, T, f_{\nu_1\nu_2})_{\nu_1 T \nu_2}$ is a coarse morass.

<u>Proof</u>: (M0),(M5) are clear. For (M1)-(M4), first note that if $\eta \in S^1$ and $\alpha = \alpha_\eta$, then:

(1): If $\tau \in S_\alpha \cap \eta$, then $\beta(\tau)+1 < \eta$ (this is clear since $L_\eta = J_\eta$, $L_\eta \models \text{card } \tau = \alpha$, but τ is a cardinal of $J_{\beta(\tau)}$),

(2): hence, for $m < \omega$, setting $\rho_m = \rho^m_{\beta(\tau)}$, $A = A^m_{\beta(\tau)}$, $(J_{\rho_m}, A_m) \in L_\eta$; in particular $\mathfrak{A}_\tau \in L_\eta$,

(3): the function $(\mathfrak{A}_\tau : \tau \in S_\alpha \cap \eta)$ is uniformly $\Sigma_1(L_\eta)$ in α.

This suffices for (M2),(M4), and (M1) is easy, since $S_\alpha \cap \eta$ is uniformly $\Sigma_1(L_\eta)$ in α, $f_{\nu_1\nu_2} : L_{\nu_1} \to_Q L_{\nu_2}$ and $f(\alpha_{\nu_1}) = \alpha_{\nu_2}$.

For (M3), I'll refer to the schema of (3.23). First, let $\alpha < \alpha_\tau$ be such that $\{\alpha_\nu : \nu T \tau\}$ is unbounded in α. Let $X = \bigcup\{\text{range } f_{\nu\tau} : \nu T\tau \text{ and } \alpha_\nu < \alpha\}$, and let $Y = X \cap L_\tau$. Then $\alpha_\tau \cap X = \alpha_\tau \cap Y = \alpha$, and $p(\tau) \in X$. Now $\mathfrak{A}_\tau|X$ is the union of a tower of Σ_1-elementary substructures of $\mathfrak{A}_\tau$ and $L_\tau|Y$ is the union of a tower of Q-elementary substructures of L_τ, and so $\mathfrak{A}_\tau|X \prec_1 \mathfrak{A}_\tau$, $L_\tau|Y \prec_Q L_\tau$. In fact, it's easy to see that $X = h_{\mathfrak{A}_\tau}''(\omega\times\alpha\times\{p(\tau)\})$. Also, $\tau < \rho(\tau)$ iff $\tau \in X$. Now let $\bar{f}: (J_{\bar{\rho}}, \bar{A}, \bar{p}) \xrightarrow{\sim} \mathfrak{A}_\tau|X$ and let $\bar{f}': L_{\bar{\tau}} \xrightarrow{\sim} L_\tau|Y$. It's easy to see that $\bar{f} \supseteq \bar{f}'$ and that if $\tau \in X$, then $\bar{f}(\bar{\tau}) = \tau$, and $\bar{\tau}$ is a regular cardinal of $J_{\bar{\rho}}$ and in fact is the cardinal successor in $J_{\bar{\rho}}$ of α which is, even without assuming $\tau \in X$, the unique uncountable cardinal of $L_{\bar{\tau}}$. Further, $\bar{\tau}$ is a limit of non-projectible ordinals: this is because $L_\tau|Y \cong L_{\bar{\tau}}$ and $L_\tau|Y$ is the union of a tower of Q-elementary substructures of L_τ in which the appropriate Π_2-properties hold. This means that $\bar{\tau} \in S^1$ and $\alpha = \alpha_{\bar{\tau}}$. To see that $\bar{\tau}T\tau$ is now easy using the argument of (3.23).

It remains only to verify that (M6),(M7) hold in order to conclude that $(S,S^0,S^1,T,f_{\nu_1\nu_2})_{\nu_1 T \nu_2}$ is a morass. This will be done in the next two lemmas where the argument of (3.23) will be combined with (2.11).

(3.35) LEMMA: (M6) holds for $(S,S^0,S^1,T,f_{\nu_1\nu_2})_{\nu_1 T\nu_2}$.

Proof: First assume $\rho(\nu_0) > \nu_0$. Let $\rho_i = \rho(\nu_i)$, $\beta_i = \beta(\nu_i)$, $A_i = A(\nu_i)$, $p_i = p(\nu_i)$ $(i = 0,2)$; let $n = n(\nu_0) = n(\nu_2)$, let $f = f_{\nu_0\nu_2}$, let $g_0 = f_{\nu_0\nu_2}|L_{\nu_0}$, and let $g_1 = id|L_{\nu_1}$. Let ρ_1, A_1, f_0, f_1 be as guaranteed by (2.11). Thus, among other things ν_1 is, in J_{ρ_1}, the cardinal successor of α_{ν_2}.

By (2.9)d) applied to f_1 there are $\beta_1, \tilde{f}_1$ such that $\tilde{f}_1 \supset f_1, \tilde{f}_1: J_{\beta_1} \to_n J_{\beta_2}$, $\rho_1 = \rho^n_{\beta_1}$, $A_1 = A^n_{\beta_1}$ and with the other properties of (2.9)d). Thus, in J_{β_1}, ν_1 is the cardinal successor of α_{ν_2}. Since $f = f_1 \circ f_0$, $p_2 \in$ range f_1, say $f_1(p_1) = p_2$; thus, also $f_0(p_0) = p_1$.

Again, apply (2.9)d), this time to f_0, to obtain $\tilde{f}_0 \supseteq f_0$ such that $\tilde{f}_0: J_{\beta_0} \to_{n+1} J_{\beta_1}$ and having the other properties of (2.9)d) (that $\tilde{f}_0$ is Σ_{n+1} - and not just Σ_n-elementary follows from the fact that f_0 is Σ_1-elementary since it's cofinal). Let $\bar{g}$ be a map of a subset of α_{ν_0} onto ν_0 which is $\Sigma_{n+1}(J_{\beta_0})$ in $q_0 \in J_{\beta_0}$, say by the Σ_{n+1}-formula θ.

Let g be the subset of J_{β_1} which is defined over J_{β_1} by θ and $q_1(=\tilde{f}_0(q_0))$. By the Σ_{n+1}-elementarity of $\tilde{f}_0$, g is a map from a subset of α_{ν_2} into ν_1, and $f_0"\bar{g} \subseteq g$. Hence g is cofinal into ν_1. But ν_1 is, in J_{β_1}, the cardinal successor of α_{ν_2}. Hence, in the usual fashion, g gives rise to a $\underset{\sim}{\Sigma}_{n+1}(J_{\beta_1})$ map of a subset of α_{ν_2} onto ν_1. Thus $\beta_1 = \beta(\nu_1)$ and $n(\nu_1) \leq n$.

If $m < n$, then $\rho^m_{\beta_1} \geq \rho_1$ and $\rho_1 > \nu_1$. Thus, if $x \subseteq J_{\nu_1}$ is $\underset{\sim}{\Sigma}_{m+1}(J_{\beta_1})$, $x \in J_{\beta_1}$, by (2.9)b). Thus, no such x can be a map of a subset of a smaller ordinal onto ν_1. This means that $n(\nu_1) \geq n$, i.e. $n(\nu_1) = n$, so that $\rho_1 = \rho(\nu_1)$, $A_1 = A(\nu_1)$. Suppose $p_1 \neq p(\nu_1)$. Let $\xi < \alpha_{\nu_2}$, $i < \omega$ be such that $p_1 = h_{\rho_1,A_1}(i,\xi,p(\nu_1))$. Let $q_2 = f_1(p(\nu_1))$. Then $p_2 = h_{\rho_2,A_2}(i,\xi,q_2)$. But then $p_2 \leq_J q_2$ so $p_1 \leq_J p(\nu_1)$. On the other hand, this means that in (J_{ρ_1},A_1) the following Σ_1-formula holds:

$$(\exists q)(\exists \xi<\alpha_{\nu_2})(\exists i<\omega)(p_1=h_{\rho_1,A_1}(i,\xi,q) \text{ and } q <_J p_1).$$

But that means that in (J_{ρ_0},A_0) the following Σ_1-formula holds:

$$(\exists q)(\exists \xi<\alpha_{\nu_0})(\exists i<\omega)(p_0=h_{\rho_0,A_0}(i,\xi q) \text{ and } q <_J p_0).$$

This, however, contradicts the defining property of $p_0 = p(\nu_0)$. Hence $p_1 = p(\nu_1)$. But then $f_0 = f_{\nu_0\nu_1}$ witnesses that $\nu_0 T \nu_1$ and, as required, $f_0|\nu_0 = g_0 = f_{\nu_0\nu_2}|\nu_0$.

If $\rho(\nu_0) = \nu_0$, then the proof proceeds essentially as above but with a few minor modifications which I now give. First, there's no need to invoke (2.11): set $\rho_1 = \nu_1$, $f_0 = f_{\nu_0\nu_2}$, $f_1 = id|J_{\rho_1}$. Let $A_1 = A_2 \cap J_{\rho_1}$. Then A_1 is also $\bigcup_{\xi<\nu_1} A_2 \cap J_\xi = \bigcup_{\xi<\nu_0} f_0(A_0 \cap J_\xi)$. Thus (J_{ρ_1},A_1) is amenable since each $f_o(A_o \cap J_\xi) \in J_{\rho_1}$. Also $(J_{\rho_1}, A_1) \prec_o (J_{\rho_2}, A_2)$. Now proceed as before to obtain $p_1, \beta_1, \tilde{f}_o, \tilde{f}_1$ as before. If $\rho_1 < \beta_1$, then ρ_1 is a cardinal of J_{β_1} by (2.3)b), and ρ_1 is regular in J_{β_1} since it is, in J_{β_1}, the cardinal successor of α_{ν_2}. The rest of the proof goes through exactly as before, but some care must be taken if $\rho_1 = \beta_1$. This is no obstacle to concluding, as before, that $\beta_1 = \beta(\nu_1)$, $n(\nu_1) \leq n$. However, if $\rho_1 = \beta_1$ another argument is needed to conclude that $n(\nu_1) = n$: if $n(\nu_1) < n$, let $m = n(\nu_1)$. Then $\rho_1 = \rho^m_{\beta_1} = \beta_1$, and so there is a $\underset{\sim}{\Sigma}_{m+1}(J_{\beta_1})$ map σ of a subset of α_{ν_2} onto β_1. Hence $\rho^{m+1}_{\beta_1} \leq \alpha_{\nu_2}$, which is impossible since $m+1 \leq n$ and so $\rho^{m+1}_{\beta_1} = \beta_1$. The proof that $p_1 = p(\nu_1)$ is as before.

(3.26) The following will be useful in the proof of (M7).

<u>PROPOSITION:</u> Suppose that $\bar{\tau}T\tau$ and $\tau = \sup f_{\bar{\tau}\tau}``\bar{\tau}$. Then $\omega\rho(\tau) = \sup f_{\bar{\tau}\tau}``\omega\rho(\bar{\tau})$.

<u>Proof:</u> If not, let $\omega\rho' = \sup f_{\bar{\tau}\tau}``\omega\rho(\bar{\tau})$. Then, setting $A' = A(\tau) \cap J_{\rho'}$, $(J_{\rho'}, A')$ is amenable, since $A' = \bigcup_{\xi<\omega\rho'} A(\tau) \cap S_\xi = \bigcup_{\xi<\omega\rho(\bar{\tau})} f_{\bar{\tau}\tau}(A(\bar{\tau}) \cap S_\xi)$ and each $f_{\bar{\tau}\tau}(A(\bar{\tau}) \cap S_\xi) \in J_{\rho'}$. Also, $(J_{\rho'}, A') \rightarrow_o (J_{\rho(\tau)}, A(\tau))$. Let $\sigma = \mathrm{id}|J_{\rho'}$. Let $n = n(\bar{\tau}) = n(\tau)$.

By (2.9)d), let $\tilde{\sigma}, \beta'$ be such that $\tilde{\sigma} \supseteq \sigma$, $\tilde{\sigma}: J_{\beta'} \rightarrow_n J_{\beta(\tau)}$, $\rho' = \rho^n_{\beta'}$, $A' = A^n_{\beta'}$ with the other properties of (2.9)d). Also, by (2.9)d) applied to $f_{\bar{\tau}\tau}$ viewed as a Σ_1-elementary map between $(J_{\rho(\bar{\tau})}, A(\bar{\tau}))$ and $(J_{\rho'}, A')$, let $\tilde{f} \supseteq f_{\bar{\tau}\tau}$, $\tilde{f}: J_{\beta(\bar{\tau})} \rightarrow_{n+1} J_{\beta'}$. Accordingly as in (3.25), there is a $\underset{\sim}{\Sigma}_{n+1}(J_{\beta'})$ map of a subset of α_τ onto τ. This means that $\beta(\tau) = \beta'$, but this is absurd, because $\beta' < \beta(\tau)$ since $\rho^n_{\beta'} = \rho' < \rho = \rho^n_{\beta(\tau)}$.

(3.27) LEMMA: (M7) holds for $(S, S^o, S^1, T, f_{\nu_1\nu_2})$ $\nu_1 T \nu_2$.

Proof: I'll prove the second formulation of (M7). Recalling the notation there, I'll first prove that there is $\nu_1 T \nu_2$ with $\alpha_{\nu_1} = \alpha_1$. This will be by cases depending on whether or not $\rho(\nu_o) = \nu_o$, the case where this is true being the easy case. Then I'll prove that this ν_1 is an immediate T-successor of ν_o.

Adopting the notation of the second formulation of (M7), for $\eta_o \in S_{\alpha_o} \cap \nu_o$, let $X_{\eta_o} = f_{\eta_o^*\eta_2}{}''L_{\eta_o^*}$. Thus $L_{\nu_2}|X_{\eta_o} \prec_o L_{\nu_2}$, and $\alpha_2 \cap X_{\eta_o} = \alpha_{\eta_o^*}$. Also $\eta_{o,o} \le \eta_{o,1} \Longrightarrow X_{\eta_{o,o}} \subseteq X_{\eta_{o,1}}$. Thus setting $X = \bigcup_{\eta_o \in S_{\alpha_o} \cap \nu_o} X_{\eta_o}$, $L_{\nu_2}|X \rightarrow_o L_{\nu_2}$, and since $\alpha_1 = \sup\{\alpha_{\eta_o^*}: \eta_o \in \nu_o\}$, $\alpha_2 \cap X = \alpha_1$, and $\alpha_2 \in X$. Also for $\eta_o \in S_{\alpha_o} \cap \nu_o$, $\eta_2 \in X$, since ν_o is a limit in S_{α_o} and if $\eta_o < \eta_o'$, then $\eta_2 \in X_{\eta_o'}$. Thus, since $\nu_2 = \sup f_{\nu_o\nu_2}{}''\nu_o = \sup\{\eta_2: \eta_o \in S_{\alpha_o} \cap \nu_o\}$, X is cofinal in ν_2 so $L_{\nu_2}|X \prec_1 L_{\nu_2}$ and in fact $L_{\nu_2}|X \rightarrow_Q L_{\nu_2}$.

Let $g_1: L_{\nu_1} \xrightarrow{\sim} L_{\nu_2}|X$. Then it is easily seen that $g_1(\alpha_1) = \alpha_2$, $g_1|\alpha_1 = \mathrm{id}|\alpha_1$, that α_1 is the unique uncountable cardinal of L_{ν_1} and that ν_1 is a limit of non-projectible ordinals; i.e. $\nu_1 \in S^1$, $\alpha_1 = \alpha_{\nu_1}$. Also, it's clear that $f_{\nu_o\nu_2}{}''L_{\nu_o} \subseteq X$, and so let $g_o = g_1^{-1} \circ f_{\nu_o\nu_2}|L_{\nu_o}: L_{\nu_o} \rightarrow_Q L_{\nu_1}$. Also, let $n = n(\nu_o) = n(\nu_2)$.

First suppose $\rho(\nu_o) > \nu_o$, $\rho(\nu_2) > \nu_2$. Recall that $\rho_i = \rho(\nu_i)$, $A_i = A(\nu_i)$, $p_i = p(\nu_i)$ $(i = 0,2)$. By (2.11) let f_o, f_1, ρ_1, A_1 be such that $\rho_1 > \nu_1$, ν_1 is a regular cardinal of J_{ρ_1} (and hence ν_1 is in J_{ρ_1} the cardinal successor of α_1),

(J_{ρ_1}, A_1) is amenable, $f_i \supseteq g_i$ $(i = 0,1)$,

$f_0: (J_{\rho_0}, A_0) \to_{0,\text{cofinal}} (J_{\rho_1}, A_1)$, $f_1: (J_{\rho_1}, A_1) \to_0 (J_{\rho_2}, A_2)$. In fact, f_1 is cofinal in $\omega\rho_2$: this is because $f_1 \circ f_0 = f_{\nu_0\nu_2}$ and by (3.26) $f_{\nu_0\nu_2}$ is cofinal in $\omega\rho_2$. But then f_1 is Σ_1 and in fact Q-elementary between (J_{ρ_1}, A_1) and (J_{ρ_2}, A_2); also $p_2 \in$ range f_1, since $p_2 \in$ range f_0, say $p_2 = f_1(p_1)$. Now, applying (3.23), $\nu_1 T \nu_2$.

If $\rho(\nu_0) = \nu_0$, $\rho(\nu_2) = \nu_2$, let $\rho_1 = \nu_1$, let $f_i = g_i$ $(i = 0,1)$ and let $A_1 = \bigcup_{\eta_0 \in S_{\alpha_0} \cap \nu_0} f_0(A_0 \cap L_{\eta_0})$; thus, as before, (J_{ρ_1}, A_1) is amenable. Also, let p_1 be such that $f_1(p_1) = p_2$. Again by (3.23), $\nu_1 T \nu_2$.

Since T is a tree, this means that $\nu_0 T \nu_1$. It remains only to see that ν_0 is the immediate T-predecessor of ν_1. Suppose that $\alpha_0 < \alpha' < \alpha_1$ and $\alpha' \in S^0$. Since $\alpha' < \alpha_1$ there is $\eta_0 \in S_{\alpha_0} \cap \nu_0$ such that $\alpha' < \alpha_{\eta_0^*}$; accordingly, there can be no $\eta_0' \in S_{\alpha'}$ such that $\eta_0 T \eta_0' T \eta_0^*$. But this also means that there can be no $\nu' \in S_{\alpha'}$ such that $\nu_0 T \nu' T \nu_1$, because, if there were such a ν', then setting $f_{\nu_0\nu'}(\eta_0) = \eta_0'$, by (M2) $\eta_0 T \eta_0'$. Also $f_{\nu_0\nu_1}(\eta_0) = f_{\nu'\nu_1}(\eta_0')$, and so, again by (M2), $\eta_0' T f_{\nu_0\nu_1}(\eta_0)$. Since T is a tree, this means $\eta_0' T \eta_0^*$ (since $\eta_0^* T f_{\nu_0\nu_1}(\eta_0)$) which is a contradiction.

(3.28) The morass just constructed has the additional property, assuming V = L, that $H_{\omega_2} = \bigcup_{\nu \in S^1} |\mathfrak{A}_\nu|$. The following definition which will be used in the next section, is an attempt to formulate an abstract version of this property.

DEFINITION: Suppose $(S,S^0,S^1,T,f_{\nu_1\nu_2})_{\nu_1 T\nu_2}$ is a κ-morass.

Suppose V is a function with domain S^1 such that

i) $V(\nu) \subseteq \nu\times\nu$,

ii) for $\nu \in S_{\alpha_\tau} \cap \tau$, $V(\nu) = V(\tau) \cap \nu\times\nu$,

iii) $U = \bigcup_{\nu\in S_\kappa} V(\nu)$ is such that $(U''\{\nu\}:\omega_1 < \nu < \omega_2)$

enumerates $[\kappa^+]^{<\kappa^+}$.

Then $(S,S^0,S^1,T,f_{\nu_1\nu_2})_{\nu_1 T\nu_2}$ is <u>universal with respect to V</u> iff whenever $\nu T\tau$, $f_{\nu\tau}|\nu$: $(\nu,\epsilon,V(\nu)) \to_0 (\tau,\epsilon,V(\tau))$.

$(S,S^o,S^1,T,f_{\nu_1\nu_2})_{\nu_1 T\nu_2}$ is <u>universal</u> iff it is universal with respect to some V satisfying i), ii), iii).

In the above construction, set $(\xi,\zeta) \in V(\nu)$ iff $\xi,\zeta < \nu$ and $\xi \in$ the $\zeta^{\underline{th}}$ (in $<_J$) subset of ν.

§4. COMBINATORIAL APPLICATIONS.

In this section I'll give some examples of combinatorial applications of morasses. Most of these were found very early, and for some it was seen, after the fact, the the full morass structure wasn't really needed. Notably absent is the application to the gap-two-two-cardinals theorem, long the principal application of morasses, but whose length and difficulty ruled out including it here. Very recently Shelah and I, [8], and Velleman [10], independently proved that the existence of morasses is equivalent to Martin's Axiom type principles, which in turn provide new and systematic methods for applying morasses. For other applications of morasses see [3] and [7].

(4.1) Recall the principle $\square_{\omega_1}$: there is a sequence $(C_\alpha$: α limit, $\alpha < \omega_2)$ such that

i) C_α is closed unbounded in α,

ii) if cf $\alpha < \omega_1$ then o.t$(C_\alpha) < \omega_1$,

iii) if β is a limit of C_α then $C_\beta = C_\alpha \cap \beta$.

The following is easily seen to be an equivalent formulation. There is a sequence $(C_\alpha : \alpha$ limit, $\alpha < \omega_2)$ such that:

i') C_α is a closed subset of the set of limit ordinals $< \alpha$ which is unbounded if cf $\alpha < \omega$,

ii'): as in ii),

iii') if $\beta \in C_\alpha$ then $C_\beta = C_\alpha \cap \beta$.

Finally, if $S \subseteq (\omega_1, \omega_2)$ is closed unbounded in ω_2, then $\square_S$ is the principle: there is a sequence $(C_\alpha : \alpha$ limit, $\alpha \in S)$ such that

i_S). C_α is a closed subset of Lim $\cap\ \alpha \cap S$ which is unbounded if cf $\alpha > \omega$, ii), and iii').

(4.2) LEMMA: If $(S, S^0, S^1, T, f_{\nu_1\nu_2})_{\nu_1 T \nu_2}$ is an ω_1-morass then $\square_{\omega_1}$ holds.

Proof: Let $\nu \in S_{\omega_1}$. Set $\tau \in C_\nu$ iff $\tau < \nu$, and there is $\bar{\nu} T \nu$ such that $\tau = \sup f_{\bar{\nu}\nu}{}''\bar{\tau}$. The easy verifications are left to the reader.

(4.3) Let $\kappa > \omega$ be regular. Recall the principle $\diamondsuit^+_\kappa$: There is a sequence $(W_\alpha : \alpha < \kappa)$ such that:

i) $W_\alpha \subseteq P(\alpha)$, card $W_\alpha < \kappa$,

ii) whenever $Y \subseteq \kappa$, there is a closed unbounded $C \subseteq \kappa$ such that for $\alpha \in C$, $Y \cap \alpha$, $C \cap \alpha \in W_\alpha$.

(4.4) LEMMA: Let κ be an infinite cardinal. If there is a universal κ^+-morass then $\diamondsuit^+_{\kappa^+}$ holds.

Proof: Let $(S,S^0,S^1,T,f_{\nu_1\nu_2})_{\nu_1 T\nu_2}$ be a κ^+-morass universal with respect to V. For $\alpha \in S^0 \cap \kappa^+$ set $a \in K_\alpha$ iff there is $\nu \in S_\alpha$, $\tau < \nu$ such that $a = \{\beta < \alpha: (\tau,\beta) \in V(\nu)\}$, and let $K'_\alpha = \{a \cap \beta: a \in K_\alpha, \beta \le \alpha\}$. Let W_α be the Boolean algebra generated by $K'_\alpha \cup \{T_\nu: \nu \in S_\alpha\} \cup \alpha$, where $T_\nu = \{\alpha_{\bar\nu}: \bar\nu T\nu\}$. For $\alpha \in \kappa^+ \setminus S^0$, let $W_\alpha = \phi$.

Now let $Y \subseteq \kappa^+$, let (ν,τ) be such that $\nu \in S_{\kappa^+}$, $\tau < \nu$, $Y = \{\beta < \kappa^+: (\tau,\beta) \in V(\nu)\}$. Set $\alpha \in C$ iff $\alpha \in T_\nu$ and letting $\bar\nu$ be that element of S_α such that $\bar\nu T\nu$, $\dot\tau \in$ range $f_{\bar\nu\nu}$. Let α_0 be the least element of C. Suppose $\alpha \in C$, and let $\nu' \in S_\alpha$ be such that $\nu' T\nu$. Then $C \cap \alpha = T_{\nu'} \setminus \alpha_0 \in W_\alpha$. Further, if τ' is such that $f_{\nu'\nu}(\tau') = \tau$, then $Y \cap \alpha = \{\beta < \alpha: (\tau',\beta) \in V(\nu')\}$, because $f_{\nu'\nu}: (\nu',\epsilon,V(\nu')) \to_0 (\nu,\epsilon,V(\nu))$ and $f_{\nu'\nu}|\alpha = id|\alpha$. But $\{\beta < \alpha: (\tau',\beta) \in V(\nu')\} \in K_\alpha \subseteq W_\alpha$.

(4.5) Let $\kappa > \omega$ be regular. Recall Silver's principle $(W)_\kappa$: there is a family $F \subseteq P(\kappa)$, card $F = \kappa^+$ such that:

(*): for $\alpha < \kappa$, setting $F|\alpha = \{X \cap \alpha: X \in F\}$, card $F|\alpha < \kappa$, and there is a sequence $(W_\alpha: \alpha < \kappa)$ such that $W_\alpha \in H_\kappa$, $W_\alpha \subseteq PP(\alpha)$, and: (**): whenever $s \subseteq F$ and card $s < \kappa$, there is $\alpha_0 < \kappa$ such that for all $\beta \ge \alpha_0$, $s|\beta = \{X \cap \beta: X \in s\} \in W_\beta$.

(4.6) LEMMA: If κ is an infinite cardinal and there is a universal κ^+-morass, then $(W)_{\kappa^+}$ holds.

Proof: If $\nu \in S^1$, let $T'_\nu = \{\bar\nu: \bar\nu T\nu\}$. Set $F = \{T'_\nu: \nu \in S_{\kappa^+}\}$. Clearly card $F = \kappa^{++}$, and for $\alpha < \kappa^+$, card $F|\alpha < \kappa^+$. If $\alpha \in S^0$, set $a \in K_\alpha$ iff $a \subseteq S^1$ and there is $\nu \in S_\alpha$, $\tau < \nu$ such that either

(1) $a = \{\eta: (\tau,\eta) \in V(\nu)\}$, or

(2) there is some $\bar\nu T\nu$, $\bar\tau < \bar\nu$ such that $a = \{f_{\bar\nu\nu}(\bar\eta): (\bar\tau,\bar\eta) \in V(\bar\nu), \bar\eta \in S^1\}$.

Let $K_\alpha^1 = \{\{T'_\eta: \eta \in a\}: a \in K_\alpha\}$; let $K_\alpha^2 = \{\{T'_\eta \cup \{\eta\}: \eta \in a\}: a \in K_\alpha\}$. Let $K_\alpha^3 = \bigcup_{\beta \in S^o \cap \alpha} K_\beta^2$, let $K_\alpha^4 = \{\{T'_\eta \cup \{\eta\}\}: \eta \in S^1 \cap \alpha\}$, and let $K_\alpha^5 = \{\{T'_\eta\}: \eta \in \bigcup_{\beta \in S^o \cap \alpha+1} S_\beta\}$. Let W_α be the Boolean algebra generated by $\bigcup_{1 \le i \le 5} K_\alpha^i$.

If $\beta \in \kappa^+ \backslash S^o$, let β^* = the least element of $S^o \backslash \beta$, and let $W_\beta = \{\{X \cap \beta: X \in A\}: A \in W_{\beta^*}\}$. Then clearly $W_\alpha \in H_{\kappa^+}$, $W_\alpha \subseteq PP(\alpha)$. For (**), suppose $s \subseteq F$, card $s \le \kappa$; say $s = \{T'_\nu: \nu \in s^1\}$. Let $(\nu_i: i < \xi)$ be the increasing enumeration of s^1. The proof goes by induction on ξ. Since each W_β contains all the $\{T'_\eta\}$ for $\eta \in S_\beta$ and all the $\{T'_\eta \cup \{\eta\}\}$ for $\eta \in S^1 \cap \beta$, suppose, without loss of generality that ξ is a limit ordinal and that (**) holds for all the $(s)^\zeta = \{T'_{\nu_i} : i < \zeta\}$, for $\zeta < \xi$, with β_ζ playing the role of the α_o of (**).

Let $\lambda = \sup_{i<\xi} \nu_i$, let $\nu^* > \lambda$ and τ^* be such that (ν^*, τ^*) is lexicographically minimal such that $s^1 = \{\nu: (\tau^*, \nu) \in V(\nu^*)\}$. Choose $\bar{\nu}^* T \nu^*$ minimal such that τ^*, λ and each $\nu_i (i<\xi) \in$ range $f_{\bar{\nu}^*\nu^*}$, and $\alpha_{\bar{\nu}^*} \ge \sup_{\zeta<\xi} \beta_\zeta$.

I'll prove that $\alpha_{\bar{\nu}^*}$ witnesses that (**) holds for s. Let $\bar{\lambda}$ be such that $f_{\bar{\nu}^*\nu^*}(\bar{\lambda}) = \lambda$, and let $\bar{\tau}^*$ be such that $f_{\bar{\nu}^*\nu^*}(\bar{\tau}^*) = \tau^*$. Clearly $\bar{s}^1 = f^{-1}_{\bar{\nu}^*\nu^*}[s^1] \in K_{\alpha_{\bar{\nu}^*}}$, since $\bar{s}^1 = \{\bar{\nu}: (\bar{\tau}^*, \bar{\nu}) \in V(\bar{\nu}^*)\}$. Hence $\{T'_{\bar{\nu}_i} : i < \xi\} = s|\alpha_{\bar{\nu}^*} \in W_{\alpha_{\bar{\nu}^*}}$. By the definition of $W_\gamma (\gamma \in \kappa^+ \backslash S^o)$ it clearly suffices to prove that $s|\beta \in W_\beta$ for $\beta \in S^o \cap \kappa^+$, $\beta > \alpha_{\bar{\nu}^*}$.

So, let $\beta > \alpha_{\bar{\nu}^*}$, $\beta \in S^o \cap \kappa^+$. If there is $\lambda' \in S_\beta$ with $\bar{\lambda} T \lambda' T \lambda$, then $f_{\bar{\lambda}\lambda'}$ "$\bar{s}^1 \in K_\beta$ and so again $s|\beta \in W_\beta$, since $s|\beta = f_{\bar{\lambda}\lambda'}$"$s|\alpha_{\bar{\nu}^*}$.

Suppose then that there are λ_1, λ_2 such that $\bar{\lambda}\underline{\underline{T}}\lambda_1 T\lambda_2 T\lambda$, λ_2 is an immediate T-successor of λ_1 and $\alpha_{\lambda_1} < \beta < \alpha_{\lambda_2}$. Let ν_i^1, ν_i^2 ($i<\xi$) be such that $f_{\lambda_1\lambda}(\nu_i^1) = \nu_i$, $f_{\lambda_2\lambda}(\nu_i^2) = \nu_i$ respectively ($i<\xi$). Thus $f_{\lambda_1\lambda_2}(\nu_i^1) = \nu_i^2$. For $i < \xi$ let η_i be the least η such that $\nu_i^1 T\eta\ T\ \nu_i^2$. By (M7), $\sup_{i<\xi} \alpha_{\eta_i} = \alpha_{\lambda_2}$. Thus, there is $\theta < \xi$ such that $\alpha_{\eta_\theta} > \beta$.

For $\theta \le i < \xi$, $T'_{\nu_i} \cap \beta = (T'_{\nu_i} \cap \alpha_{\lambda_1}) \cup \{\nu_i^1\} = T'_{\nu_i^1} \cup \{\nu_i^1\}$. But $\{T'_{\nu_i^1} \cup \{\nu_1^1\}: \theta \le i < \xi\} \in K^2_{\lambda_1} \subseteq W_\beta$. But $s|\beta =$ $\{T'_{\nu_i} \cap \beta: i < \theta\} \cup \{T'_{\nu_i} \cup \beta: \theta \le i < \xi\}$, and, by the induction hypotheses $\{T'_{\nu_i} \cap \beta: i < \theta\} = (s)^\theta|\beta \in W_\beta$. Since W_β is closed under finite unions, $s|\beta \in W_\beta$.

§5. GETTING MORASSES BY FORCING.

My approach to adding a $(\kappa,1)$ morass is to add a somewhat simpler object which is then "thinned" to obtain a morass. This is essentially as in Jensen's original treatment, see [9]. However my treatment here is based on a new set of conditions introduced by Velleman, see [10], who has kindly permitted me to present them here and in [8]. The basic difference is that Jensen's conditions involved first adding a $\square_\kappa$-sequence and then picking out a subtree of a tree with maps along it which is canonically associated with the $\square_\kappa$ sequence, whereas Velleman's conditions permit great latitude in simultaneously assembling the tree and the maps.

First, I'll define the generic object.

(5.1) α and β are the <u>same kind</u> of ordinal if they're both zero, or both non-zero limits, or both successors. $f: \alpha \to \beta$ is <u>nice</u> iff f is order-preserving, α and β are the same kind of ordinals, for all $\gamma < \alpha$, γ and $f(\gamma)$ are the same kind of ordinals and if $\gamma+1 < \alpha$ then $f(\gamma+1) = f(\gamma)+1$ and, finally, if $\alpha = \bar{\alpha}+1$, then $\beta = f(\bar{\alpha})+1$.

DEFINITION: Let $\kappa \geq \omega_1$ be regular. A $(\kappa,1)$ premorass is a triple $(T, \rightarrow, f_{xy})_{x \prec y}$ such that

i) $(T, \rightarrow)$ is a tree, $\{\kappa\}\times\kappa^+ \subseteq T \subseteq \kappa+1\times\kappa^+$, $(\mathrm{dom}\ T)\cap\kappa = \kappa$, and for $\alpha < \kappa$, $0 < \gamma_\alpha = \{\tau : (\alpha,\tau) \in T\} < \kappa$ is a limit,

ii) for $x \in T$, set $x = (\ell(x), \mathcal{O}(x))$ (warning: $\ell(x)$ need not coincide with the level of x in $(T, \rightarrow)$); then $(f_{xy} : x \rightarrow y)$ is a commutative system of nice maps $f_{xy} : \mathcal{O}(x) \to \mathcal{O}(y)$,

iii) if $x \rightarrow y$, $\tau < \mathcal{O}(x)$, $w = (\ell(x),\tau)$, $z = (\ell(y), f_{xy}(\tau))$, then $w \rightarrow z$ and $f_{wz} = f_{xy}|\tau$.

iv') $\{\ell(x) : x \rightarrow y\}$ is either empty or a coinitial segment of $\mathrm{dom}\ T \cap \ell(y)$, and is non-empty if $\ell(y)$ is limit,

v') if $\ell(y)$ is limit then $\mathcal{O}(y) = \bigcup_{x \rightarrow y} \mathrm{range}\ f_{xy}$,

vi') if $\mathcal{O}(x)$ is limit, $x \rightarrow y$ and $\lambda = \sup \mathrm{range}\ f_{xy} < \mathcal{O}(y)$ then setting $z = (\ell(y),\lambda)$, $x \rightarrow z$, $f_{xz} = f_{xy}$.

The next lemma justifies the introduction of premorasses by showing that they can be "thinned" to yield morasses.

LEMMA (Jensen, essentially): If there is a $(\kappa,1)$ premorass there is a $(\kappa,1)$-morass.

Proof: To fix ideas, take $\kappa = \omega_1$. As usual, the generalization is straightforward. First, define a normal sequence $(\beta_\xi : \xi < \omega_1)$ of countable ordinals by induction: $\beta_0 = 0$, $\beta_\lambda = \sup_{\xi<\lambda} \beta_\xi$, $\beta_{\xi+1}$ is the least $\beta > \beta_\xi$ such that $\gamma_\eta < \beta$ for all $\eta < \beta$. Set $S^0 = \{\beta_\xi : \xi < \omega_1\}$.

If $x \in T$, set $\nu(x) = \ell(x)+\mathcal{O}(x)$, $\alpha(x) = \ell(x)$. If $x \prec y$, let $\bar{f}_{xy}: \nu(x)+1 \to \nu(y)+1$ be such that $\bar{f}_{xy}|\alpha(x) = \mathrm{id}|\alpha(x)$, for $\xi < \mathcal{O}(x)$, $\bar{f}_{xy}(\alpha(x)+\xi) = \alpha(y)+f_{xy}(\xi)$, $\bar{f}_{xy}(\nu(x)) = \nu(y)$. Then $\bar{f}_{xy}$ is nice. Set $\nu(x) \prec \nu(y)$ iff $x \prec y$.

If $x \in T$, $\alpha(x) \in S^0$, $\nu = \nu(x)$ let

$D_\nu = \{0\}\times\{(\beta,\gamma_\beta): \beta < \alpha(x)\} \cup \{1\}\times(S^0 \cap \nu(x)) \cup$

$\{2\}\times\{(\nu(y), \nu(z),\zeta,\xi): y \prec z,\ \nu(z) < \nu,\ \bar{f}_{yz}(\zeta) = \xi\}$. Thus $D^+_\nu \subseteq 3 \times^{\omega} \nu$ codes up the segment of the premorass internal to ν. Further, if $\nu < \nu(w)$ and $\alpha(w) \in S^0$, then $D_\nu = D_{\nu(w)} \cap 3 \times^{\omega} \nu$. ν is <u>D-admissible</u> iff $\mathfrak{A}_\nu = (L_\nu[D_\nu]$ is admissible.

Set $\nu(x) \in S^1$ iff $x \in T$, $\alpha(x) \in S^0$, $\nu(x)$ is a limit of D-admissibles, and $\mathfrak{A}_{\nu(x)} \models \alpha(x) = \aleph_1$. $S = \{(\alpha(x),\nu(x)) : \nu(x) \in S^1\}$. Note that all elements of $L_\nu[D_\nu]$ are $\Sigma_1(\mathfrak{A}_\nu)$-definable in parameters from α_ν.

For $\nu,\tau \in S^1$, set $\nu \prec^+ \tau$ iff $\nu \prec \tau$ and there is $f^+ \supseteq \bar{f}_{\nu\tau}$, $f^+: \mathfrak{A}_\nu \to_Q \mathfrak{A}_\tau$. Such a map is unique if it exists, in which case, call it $f^+_{\nu\tau}$.

<u>CLAIM</u>: $(S, S^0, S^1, \prec^+, \bar{f}_{\nu\tau})_{\nu \prec^+ \tau}$ is a morass.

<u>Proof of CLAIM</u> (and thus of lemma): (M1) is satisfied because of the Q-elementarity of $f^+_{\nu\tau}$. (M2 is satisfied since by the properties of the premorass $\bar{\nu} \prec \nu$, and because $f^+_{\bar{\tau}\tau}(\mathfrak{A}_{\bar{\nu}}) = (\mathfrak{A}_\nu)$ this means $f^+_{\bar{\tau}\tau}|L_{\bar{\nu}}[D_{\bar{\nu}}]: \mathfrak{A}_{\bar{\nu}} \to_\omega \mathfrak{A}_\nu$. (M3) follows from the properties of the premorass and easy model-theoretic arguments.

(M4) uses the fact that if $\eta \in S_{\alpha_\tau}$ and $\eta > \tau$ then since $\mathfrak{A}_\eta \models \alpha_\tau$ is regular, a familiar Lowenheim-Skolem-type argument for $\mathfrak{A}_\tau$ can be carried out within $\mathfrak{A}_\eta$ to give an elementary tower of height α_τ of submodules of $(\mathfrak{A}_\tau, D^+_\tau)$ whose intersections with α_τ form a closed-unbounded subset of $S^0 \cap \alpha_\tau$.

Thus, among other things, α_τ is a limit in S^0, and so τ is non-trivial in $\prec$. The conclusion now follows readily.

(M5) is trivially verified. (M6) follows from the corresponding property vi) of the premorass, the fact that $f^+_{\bar{\tau}\tau}: \mathcal{O}\!\iota_{\bar{\tau}} \to_0 \mathcal{O}\!\iota_\lambda$, since $\mathcal{O}\!\iota_\lambda \prec_0 \mathcal{O}\!\iota_\tau$, and the fact that $f^+_{\bar{\tau}\tau}$ is cofinal in λ.

For (M7), first, α can be taken, without loss of generality with $\alpha_{\bar{\tau}} < \alpha < \alpha_\tau$. Then, by the property iv') of the premorass, there is $\tau' \in S_\alpha$ such that $\bar{\tau} \prec \tau' \prec \tau$. For $\bar{\nu} \in S_{\alpha_{\bar{\tau}}} \cap \bar{\tau}$, let $\nu = \bar{f}_{\bar{\tau}\tau}(\bar{\nu})$ and let ν' be that element of S_α such that $\bar{\nu} \prec^+ \nu' \prec^+ \nu$. Let $g = \bigcup_{\bar{\nu} \in S_{\alpha_{\bar{\tau}}} \cap \bar{\tau}} f^+_{\nu'\nu}$. Then, by arguments similar to those for (M6), $g: \mathcal{O}\!\iota_{\tau'} \to_Q \mathcal{O}\!\iota_\tau$. But then $\tau' \prec^+ \tau$ and $g = f^+_{\tau'\tau}$. This completes the proof of the CLAIM and the LEMMA.

I turn now to presenting Velleman's conditions for adding a premorass.

(5.3) <u>DEFINITION:</u> $p = (t, \prec, f_{xy})_{x \prec y} \in P$ iff

i) $t \subseteq (\kappa+1) \times \kappa^+$, card $t < \kappa$, $t \neq \emptyset \Rightarrow ((\kappa,0) \in t$ and dom $t \cap \kappa$ is a successor ordinal), $(t, \prec)$ is a tree, and for $\alpha \in \text{dom } t \cap \kappa$, $0 < \gamma^p_\alpha = \{\tau : (\alpha,\tau) \in t\} < \kappa$ is a limit,

ii): ii), iii), iv'), vi) of (5.1) hold, replacing T by t (note that for iii) this means that if $\ell(y) = \kappa$, then $(\kappa, f_{xy}(\tau)) \in t$ and for vi) if $\ell(y) = \kappa$, then $(\kappa,\lambda) \in t$),

(iii): v') of (5.1) holds with the additional requirement that $\ell(y) < \kappa$. For $p \in P$, $p = (t^p, \prec^p, f^p_{xy})_{x \prec^p y}$, and $t^p = \bigcup_{\alpha \in (\text{dom } t^p) \cap \kappa} (\{\alpha\} \times \gamma^p_\alpha) \cup \{\kappa\} \times S^p_\kappa$. Then, for $p,q \in P$, $p \leq q$ iff

(iv) $(t^q, \prec^q)$ is a subtree of $(t^p, \prec^p)$ and if $x \prec^q y$ then $f^p_{xy} = f^q_{xy}$; for $\delta \in (\mathrm{dom}\ t^q) \cap \kappa$, $\gamma^q_\delta = \gamma^p_\delta$ and if $\tau < \gamma^q_\delta$, $x \in t^q$ then $x \prec^p(\delta,\tau)$ iff $x \prec^q(\delta,\tau)$. $\mathbb{P} = (P, \leq)$.

The following lemma summarizes the important properties of $\mathbb{P}$. The reader can either carry out the verifications (straightforward, except possibly for c)) or consult [8] where $\mathbb{P}$ is presented in greater detail.

(5.4) LEMMA: a) $\mathbb{P}$ is κ-complete (i.e. every decreasing sequence of length $< \kappa$ has a (not necessarily greatest) lower bound),

b) if $p \in P$, $\alpha < \kappa$, $\beta < \kappa^+$, $\gamma < \beta$ there is $q \leq p$ such that $\alpha \in \mathrm{dom}\ t^q$, $\beta \in S^q_\kappa$ and $\gamma \in \bigcup_{x \prec^q y} \mathrm{range}\ f^q_{xy}$, where $y = (\kappa,\beta)$,

c) $\mathbb{P}$ has the κ^+-c.c.

This immediately gives

(5.5) LEMMA: In $V^{\mathbb{P}}$ there is a $(\kappa,1)$ premorass.

(5.6) REMARKS: a) If $A \subseteq \omega_2$ (or, in the general case, if $A \subseteq \kappa^+$), the entire construction of (5.2) can be relativized to A. The approach is to start from the models $\mathfrak{A}_{\nu(x)} = (L_{\nu(x)}[D_{\nu(x)}, A \cap \nu(x)], \in, D_{\nu(x)}, A \cap \nu(x))$ with $\ell(x) = \omega_1$. Attention is restricted to transitivizations of Q-elementary substructures of the $\mathfrak{A}_{\nu(x)}$ having the form $(L_{\nu(y)}[D_{\nu(y)}, \bar{A}], \in, D_{\nu(y)}, \bar{A})$ with $y \prec x$ and the inverse of the transitivizing map extending $\bar{f}_{yx}$. In particular if $2^{\aleph_0} = \aleph_1$ and $2^{\aleph_1} = \aleph_2$ (or in the general case $2^{<\kappa} = \kappa$ and $2^\kappa = \kappa^+$) and A is chosen such that $A \cap \omega_1$ codes up H_{ω_1} and $A\cap[\omega_1,\omega_2)$ codes up H_{ω_2}, then the morass which results from the thinning of (5.2) relativized to A is universal.

b) Roughly speaking, the idea of (5.4)c) is as follows. P can be partitioned into κ equivalence classes, the content of the equivalence relation being, roughly, that p and q are equivalent just in case p,q have the same underlying structure and differ only in that S_κ^p, S_κ^q do not coincide (though they will have the same order type - this will be part of the meaning of having the same underlying structure). Then, if $X \subseteq P$ has power κ^+, there is $Y \subseteq X$ of power κ^+ coming from the same equivalence class. But then there is $Z \subseteq Y$ of power κ^+ such that $\{S_\kappa^p : p \in Z\}$ forms a Δ-system, i.e., there is a such that whenever $p,q \in Z$, $p \neq q$ implies $S_\kappa^p \cap S_\kappa^q = a$ which is an initial segment of S_κ^p, S_κ^q, and either all members of $S_\kappa^p\backslash a$ are less than all members of $S_\kappa^q\backslash a$, or conversely. The definition of the equivalence relation will be such as to guarantee that two equivalent conditions whose S_κ's have the Δ-property (i.e. the property which defines a Δ-system) are compatible. Accordingly, the elements of Z are pairwise compatible, and so X could not have been an antichain.

REFERENCES

[1] DEVLIN, K.J., ON HEREDITARILY SEPARABLE HAUSDORFF SPACES IN THE CONSTRUCTIBLE UNIVERSE, FUND.MATH. Vol.82 (1974) N$\underline{o}$ 1, pp.1-10.

[2] DEVLIN, K.J., ASPECTS OF CONSTRUCTIBILITY, LECTURE NOTES IN MATH. Vol.354 (Springer, Berlin, 1973).

[3] DEVLIN, K.J., to appear.

[4] DEVLIN, K.J., and JENSEN, R.B., MARGINALIA TO A THEOREM OF SILVER, IN LECTURE NOTES IN MATH. 499, (Springer, Berlin, 1975).

[5] DODD, A.J., and JENSEN, R.B., THE CORE MODEL, ANNALS MATH. LOGIC 20 (1981) 43-75.

[6] JENSEN, R.B., THE FINE STRUCTURE OF THE CONSTRUCTIBLE HIERARCHY, ANNALS MATH. LOGIC 4 (1972) 229-308.

[7] REBHOLZ, J., SOME CONSEQUENCES OF THE MORASS AND DIAMOND, ANNALS MATH. LOGIC 7 (1975) 361-386.

[8] SHELAH, S., and STANLEY, L., S-FORCING AND MORASSES I. A "BLACK BOX" THEOREM FOR MORASSES, WITH APPLICATIONS: SUPER-SOUSLIN TREES AND GENERALIZED MARTIN'S AXIOM, to appear.

[9] STANLEY, L., CHARACTERIZING WEAK COMPACTNESS, to appear.

[10] VELLEMAN, D., Ph.D. THESIS, UNIVERSITY OF WISCONSIN, 1980.

LIST OF PARTICIPANTS

Abraham, U., (formerly Avraham, V.) Department of Mathematics, The Hebrew University, Jerusalem, Israel.

Ambos, K., Mathematisches Instituts, Theresienstr., D-8000 München, BRD.

Andersen, J., Crewe & Alsager College, Alsager, Stoke-on-Trent.

Apter, A., Department of Mathematics, M.I.T., Cambridge, MA 02139, US.

Arens, Y., Department of Mathematics, Evans Hall, U.C.B., Berkeley, CA 94720, U.S.

Baumgartner, J., Department of Mathematics, Dartmouth College, Hanover, New Hampshire 03755, U.S.

Blass, A., Department of Mathematics, University of Michigan, Ann Arbor, Michigan 48109, U.S.

Brown, M., Peterhouse, Cambridge.

Carlson, T., Department of Mathematics, University of Colorado, Boulder, Colorado, U.S.

Carr, D.M., Department of Mathematics, McMaster University, Hamilton, Ontario L8S 4K1.

Dahlhaus, S., Osthofener Weg 17, 1000 BLN-W 38, W. Germany.

Dodd, A. New College, Oxford.

Donder, H-D., Seminar für Logik, 53000 Bonn, Beringstr. 6, W. Germany.

Drake, F., School of Mathematics, The University, Leeds LS2 9JT.

Franek, F., 66 Spadina, Apart. 904, Toronto, Canada.

Erdös, P., Mathematical Institute, Hungarian Academy of Science, Realtanoda u 13-15, Budapest V, Hungary.

Falaki, M., Wolfson College, Oxford.

Farrington, P., School of Mathematics, The University of Leeds LS2 9JT.

Ferro, R., Via Gabelli 57, 35100 Padua, Italy.

Friedman, S., Department of Mathematics, U.C.B. Berkeley, CA 94720, U.S.

Gardiner, A., Department of Pure Mathematics, University of Birmingham, PO Box 363, Birmingham B15, 2TT.

Gilow, C., Department of Mathematics, Cornell University, Ithaca, New York, U.S.

Gold, A., University of Windsor, Ontario, Canada N9B 3(4.

Griffor, E., Department of Mathematics 2-247, M.I.T., Cambridge, MA 02139, U.S.

Groszek, M., Department of Mathematics, Harvard, Cambridge MA 02138, U.S.

Guaspari, D., St. John's College, Annapolis, Maryland, U.S. 21404.

Hajnal, A., Mathematical Institute, Hungarian Academy of Science, Realtanoda u 13-15, Budapest V, Hungary.

Halpern, J.D., Department of Mathematics, University of Alabama, Birmingham AL5243.

Harrington, L., Department of Mathematics, U.C.B., Berkeley, CA 94720, U.S.

Huber-Dyson, V University of Calgary, Department of Philosophy, Calgary, Alberta T2N 14N, Canada.
Hyland, M., King's College, Cambridge.
Isbell, J., Department of Mathematics, SUNY Buffalo, U.S.
Izouvaros, A., Department of Mathematics, University of Thessalonika, Greece.
Jenkins, J., Peterhouse, Cambridge, CB2 1QU.
Johnstone, P., Department of Pure Mathematics, 16 Mill Lane, Cambridge.
Jones, M., New Hall, Cambridge.
Kanamori, A., Department of Mathematics, Harvard, Cambridge, MA 02138, U.S.
Kastanas, I., Department of Mathematics, Caltech, Pasadena, CA 91109, U.S.
Kechris, A., Department of Mathematics, Institute of Technology, Pasadena,CA 91109, U.S.
Kimmel, K., Forschungsinstitut für Anthropotechnik, 5309 Mechenheim, Buschstrasse, BRD.
Koepke, P., Department of Mathematics, U.C.B., Berkeley, CA 94720, U.S.
Koppelberg, B., Mathematical Institute, FU Berlin, Königin-Luise-Str. 24, 1000 Berlin (West) 33.
Legrand, M., Penn State University, University Park, Penn. 16801, U.S.
Levinski, J-P., Université Paris VII, U.E.R. de Mathématiques, 2 Place Jussieu, 75221 Paris, France.
Libert, D. Department of Mathematics, U.C.B. Berkeley, CA 94720, U.S.
Lin, C., Department of Mathematics, 2-270, M.I.T., Cambridge, MA 02139, U.S.
Lindström, I., Department of Philosophy, Stanford University, Stanford, California 94305, U.S.
Linton, F., Department of Mathematics, Wesleyan University, Middletown CT 06457, U.S.
Maass, W., Mathematisches Institut der Universität, Theresienstr. 39, D-8000 München 2, W. Germany.
Mackenzie, K., Max Rayne House, 109 Camden Road, London NW1.
Magidor, M., Department of Mathematics, The Hebrew University, Jerusalem, Israel.
Martin, D.A., Department of Mathematics, UCLA, Los Angeles, U.S.
Mathias, A.R.D., Peterhouse, Cambridge.
Mignone, R., Department of Mathematics, Penn State University, Univesity Park, PA 16802, U.S.
Miller, A.W., Department of Mathematics, University of Wisconsin, Madison, Wisconsin 53706, U.S.
Milner, E., Department of Mathematics & Statistics, Faculty of Science, University of Calgary, 2920 24 Ave N.N., Calgary, Canada.
Mitchell, R., The School of Mathematics, The University, Leeds LS2 9JT.
Mitchell, W., 420 E 70th St., Apartment 3A, New York City, NY 10021, U.S.
Moerdijk, I., Roeterstraat 15, Amsterdam, Netherlands.

Mosbach, M., Berthold-Schwarz-Str. 25, D-6700 Ludwigshafen/RH. 14, BRD.
Moss, J., 1 Colosseum Terrace, Albany St., London NW1.
Normann, D., Institute of Mathematics, Boks 1053, University of Oslo, Blindern, Oslo 3, Norway.
Normann, S., Institute of Mathematics & Statistics, Boks 35, Agricultural University of Norway, Ass-NLH, Norway.
Pearce, J., Department of Mathematics, UCB., Berkeley, CA 94720, U.S.
Pelletier, D., Department of Mathematics, York University, Toronto M3J 1P3, Canada.
Philp, B., Pure Mathematics Department, University of Birmingham, Edgbaston, Birmingham 15.
Prikry, K., Department of Mathematics, University of Minnesota, Minneapolis, Minnesota 55455.
Quinsey, J., Mathematical Institute, 24/29 St. Giles, Oxford.
Rado, R., 14 Glebe Road, Reading RG2 7AG.
Rothacker, E., Steinkuhlstr. 22, D-463 Bochum 1, W. Germany.
Labib Sami, R., Department of Mathematics, Faculty of Science, Cairo University, Cairo, Egypt.
Shelah, S., Department of Mathematics, The Hebrew University, Jerusalem Israel.
Shore, R., Department of Mathematics, Cornell University, Ithaca, NY 14853, U.S.
Silver, J., Department of Mathematics, UCB., Berkeley, CA 94720, U.S.
Singh, D., Department of Mathematics, College of Science, University of Basrah, Basrah, Iraq.
Slaman, T., Department of Mathematics, Harvard, Cambridge, MA 02138, U.S.
Smith, B., Department of Pure Mathematics, University of Cambridge.
Smorynski, C., Mathematisch Insitut der Rijksuniversiteit te Utrecht, Budapestlaan 6, Utrecht, Netherlands.
Stanley, L., Université de Clermont, Complexe Scientifique des Ceseaux, Mathematiques Pures, B.P. 45-63170, Aubière, France.
Stekeler, P., Rheingutstr. 12, D775, Konstanz, W. Germany.
Stoltenberg-Hansen, H., Institute of Mathematics, Boks 1053, University of Oslo, Blindern, Oslo 3, Nowary.
Tall, F., Department of Mathematics, University of Toronto, Ontario, Canada.
Tavares, J., Instituto Superior de Engenharia de Coimbra, Quinta da Nova, Coimbra, Portugal.
Taylor, R., Department of Philosophy, Columbia University, New York City, U.S.
Thiele, F., Breisgauer Str. 30, D1000 Berlin 38, W. Germany.
Thomason, A., Peterhouse, Cambridge.
Watson, S., Department of Mathematics, University of Toronto, Ontario, Canada.
Welch, P., Mathematical Institute, 24/29 St. Giles, Oxford.
Wolfsdorf, K., Grossbeeren Str., 78, D-1 Berlin 61, Germany.
Wong, H., Room L, Department of Mathematics, Bedford College, Regent's Park, London.
Ziegler, M., Dietrich-Schafer-Weg 38, 1000 Berlin (West) 41.